SIMON

THE
OZONE
WAR

LYDIA DOTTO and HAROLD SCHIFF

DOUBLEDAY & COMPANY, INC.
GARDEN CITY, NEW YORK
1978

Library of Congress Cataloging in Publication Data

Schiff, Harold.
The ozone war.

Includes index.
1. Atmospheric ozone.   2. Ozone—Environmental aspects.   I. Dotto, Lydia, joint author.   II. Title.
QC879.7.S3   363.6
ISBN: 0-385-12927-0
Library of Congress Catalog Card Number 77–12876

For Ralph and Arthur

ACKNOWLEDGMENTS

We would like to thank the many scientists, journalists, and industry representatives who provide this book with a cast of thousands for giving freely of their time in personal interviews. Special thanks to John Noxon and Jeremy Bernstein for encouragement and assistance at a crucial time, and to Ian Clark, Sherry Rowland and Will Kellogg for written material.

We would also like to thank Elsie Stern, Tom Hyman, and our editor Susan Schwartz.

CONTENTS

LIST OF NAMES

ABPLANALP, Robert: president, Precision Valve Corporation

ALLEN, William: chairman of the Board, The Boeing Company

ANDERSON, James: physicist, University of Michigan

BATES, David: head of the Department of Applied Mathematics, University of Belfast, Northern Ireland

BOWER, Frank: manager, Freon Products Laboratory, E. I. Du Pont de Nemours & Company; chairman, Fluorocarbon Technical Panel, Manufacturing Chemists Association

BRODERICK, Tony: director, High Altitude Pollution Program, Department of Transportation

BUMPERS, Dale: chairman, Subcommittee on the Upper Atmosphere, Senate Committee on Aeronautical and Space Sciences

CAIRNS, T. L.: director, Central Research and Development Department, Du Pont

CANNON, Robert: former assistant secretary of the Department of Transportation, in charge of CIAP

CHANG, Julius: atmospheric modeler, Lawrence Livermore Laboratory, University of California

CICERONE, Ralph: physicist, University of Michigan

CLYNE, Michael: chemist, St. Mary's College, University of London

COLEMAN, William: former Secretary of Transportation

COOKE, Bob: science writer, Boston *Globe*

COSTLE, Douglas: administrator, Environmental Protection Agency

CRUTZEN, Paul: atmospheric modeler; director, Upper Atmospheric Project, National Center for Atmospheric Research

DAVID, Edward: former science adviser to President Nixon

DAVIS, Donald: editor, *Drug and Cosmetic Industry* magazine

DICKINSON, Jack: vice president, consumer affairs, Gillette Company; former chairman, Council on Atmospheric Sciences (COAS)

DONAHUE, Tom: atmospheric scientist, University of Michigan

ENGEL, Ralph: executive director, Chemical Specialties Manufacturers Association

FLETCHER, James: former administrator, National Aeronautics and Space Administration

GOLDBURG, Arnold: former chief SST scientist, The Boeing Company

GREENWOOD, Ron: physicist, NASA-Goddard Space Flight Center

GROBECKER, Alan: project manager, Climatic Impact Assessment Program

GUTOWSKY, Herb: director, School of Chemical Sciences, University of Illinois; chairman, National Academy of Sciences Panel on Atmospheric Chemistry

HAMMOND, Allen: staff writer, *Science* magazine

HAMPSON, John: former director, Electronics Division, Canadian Armaments Research and Development Establishment

HANDLER, Philip: president, National Academy of Sciences

HARRISON, Halstead: former member of the Scientific Research Laboratory of The Boeing Company

HENRIQUES, Fred: physicist, former head of the Department of Commerce Technical Advisory Board (CTAB) on the SST

HIRSCHFELDER, Joseph: theoretical chemist, University of Wisconsin: member of CTAB

HOTZ, Robert: publisher-editor, *Aviation Week and Space Technology* magazine

HOWARD, Carl: chemist, Aeronomy Laboratory, National Oceanic and Atmospheric Administration, Boulder

HUDSON, Bob: physicist, NASA-Goddard Space Flight Center

HUNTEN, Don: physicist, University of Arizona

IKLÉ, Fred: director, Arms Control and Disarmament Agency

JOHNSTON, Harold: chemist, University of California, Berkeley

JORGENSON, Barbara: media-relations director, National Academy of Sciences

KANTROWITZ, Arthur: chairman, Avco Everett Research Laboratory

KAUFMAN, Fred: chemist, University of Pittsburgh

KELLOGG, Will: associate director, National Center for Atmospheric Research

KENNEDY, Donald: administrator, Food and Drug Administration

KOLB, Charles: physical chemist, Aerodyne Research, Inc.

LAWRENCE, Gus: director of sales, Freon Products Division, Du Pont; vice chairman, COAS

LAZRUS, Al: chemist, National Center for Atmospheric Research

LEWIS, Howard: information director, National Academy of Sciences

LODGE, James: scientific adviser, COAS

LOVELOCK, James: University of Reading, England

MACHTA, Lester: director, Air Quality Laboratory, National Oceanic and Atmospheric Administration, Maryland

MAGRUDER, William: former director, SST project, Department of Transportation

McCARTHY, Ray: laboratory director, Freon Products Laboratory, Du Pont

McCONNELL, Jack: atmospheric physicist, York University, Toronto

McDONALD, James: atmospheric physicist, University of Arizona

McELROY, Mike: atmospheric physicist, Harvard University

MOLINA, Mario: chemist, University of California, Irvine

MOSS, Frank: former senator from Utah; former chairman of the Senate Committee on Aeronautical and Space Sciences

MURCRAY, David: physicist, University of Denver

NICOLET, Marcel: aeronomer; former director, Institut d'Aeronomie Spatiale de Belgique, Brussels

PANOFSKY, Hans: meteorologist, Pennsylvania State University; member, NAS Committee on Impacts of Stratospheric Change

PITTLE, David: commissioner, Consumer Product Safety Commission

PROXMIRE, William: senator from Wisconsin

RAMANATHAN, Veerhabadrhan: atmospheric physicist, National Center for Atmospheric Research; formerly with the National Aeronautics and Space Administration

RANDALL, William: chairman, Subcommittee on Government Activities and Transportation, House Committee on Government Operations

RASOOL, Ichtiaque: deputy associate administrator, space sciences National Aeronautics and Space Administration

ROWLAND, Sherry: chemist, University of California, Irvine

SCHMELTEKOFT, Art: director, Measurements Branch, Aeronomy Laboratory, National Oceanic and Atmospheric Administration, Boulder

SCHMIDT, Alexander: former administrator, Food and Drug Administration

SCHUYLER, Roy: vice president and general manager, Organic Chemicals Department, Du Pont

SCORER, Richard: professor of theoretical mechanics, Imperial College of Science and Technology, London

SHURKIN, Joel: science writer, Philadelphia *Inquirer*

SINGER, Fred: former chairman, SST Advisory Committee, Department of Transportation

SMITH, Dorothy: news manager, American Chemical Society

SPINDEL, Bill: executive secretary, Chemistry Division, National Academy of Sciences

STOLARSKI, Richard: physicist, NASA-Goddard Space Flight Center; formerly at the University of Michigan

SULLIVAN, Walter: science writer, New York *Times*

SWIHART, John: chief SST engineer, The Boeing Company

TALLEY, Wilson: assistant administrator, Environmental Protection Agency

TRAIN, Russell: former head, Environmental Protection Agency

TUKEY, John: statistician, Bell Laboratories, Princeton University; chairman, Committee on Impacts of Stratospheric Change, National Academy of Sciences

WOFSY, Steve: atmospheric physicist, Harvard University

WOLFF, Lester: congressman from New York

LIST OF ORGANIZATIONS

AEB: Aerosol Education Bureau
CEQ: Council on Environmental Quality
CIAP: Climatic Impact Assessment Program
CISC: Committee on Impacts of Stratospheric Change
COAS: Council on Atmospheric Sciences (Industry)
CPSC: Consumer Product Safety Commission
CSMA: Chemical Specialties Manufacturers Association
CTAB: Commerce Technical Advisory Board
EPA: Environmental Protection Agency
FDA: Food and Drug Administration
IAGA: International Association of Geomagnetism and Aeronomy
IAMAP: International Association of Meteorology and Atmospheric Physics
ICAS: Interdepartmental Committee for Atmospheric Sciences
IMOS: (Ad Hoc Federal Interagency Task Force on the) Inadvertent Modification of the Stratosphere
IUGG: International Union of Geodesy and Geophysics
MCA: Manufacturing Chemists Association
NAS: National Academy of Sciences
NASA: National Aeronautics and Space Administration
NBS: National Bureau of Standards
NCAR: National Center for Atmospheric Research
NOAA: National Oceanic and Atmospheric Administration
NRDC: Natural Resources Defense Council
NSF: National Science Foundation
WAIB: Western Aerosol Information Bureau

THE OZONE WAR

PROLOGUE

There is a certain indignity—if perhaps some poetic justice—in the possibility that human beings could seriously threaten all life on earth through the use of deodorants and hair sprays.

Yet the Spray-can War, despite its macabre zaniness, or maybe because of it, was one of the most significant environmental clashes of recent times—and certainly one of the most spirited.

It had all the proper ingredients of an environmental thriller—warnings of impending disaster, a multibillion-dollar industry at stake, scientific dissension, and (the unsurprising contribution of a bureaucratic society) a plethora of committees. Add just a touch of bandwagon-jumping by federal agencies, legislators, scientific organizations, and research teams, industry associations, environmental and consumer groups, the media, and all levels of government, and you've got the hottest environmental/political show in town.

The essence of the problem can be briefly put: Chemicals called fluorocarbons, once widely used as propellants in many spray cans (and still extensively used as refrigerants) gradually percolate to the earth's upper atmosphere, to a region known as the stratosphere, where they are broken down by the sun's energy to produce chlorine atoms. The stratosphere happens to contain the protective ozone shield, and chlorine atoms chew up ozone in a singularly pernicious fashion. Ozone is a rare form of oxygen which, though poisonous to humans at ground level, is possessed of some useful qualities in the upper atmosphere, mainly an ability to absorb the sun's deadly ultraviolet radiation, preventing most of it from reaching the earth's surface. This radiation, the cause of suntans when taken in small doses is, at greater strengths, lethal to all living things. It is no coincidence that life arose on the land masses of the earth several billion years ago only after the ozone layer materialized in the stratosphere.

The potential consequences of a damaged ozone shield are rather grim. They include increases in the incidence of skin cancer; genetic mutations; damage to animals and plants (including crops), with a consequent upset in the earth's ecosystems; and global climatic changes. The extreme case—destruction of the ozone layer—would mean the end of all life on earth.

Scientists cannot predict precisely what will happen, or when, or how severe it will be. In fact they are more worried about what they don't know than what they do know about the consequences of a depleted ozone shield.

It was only at the beginning of this decade that they began to realize that human activities could alter the amount of stratospheric ozone. The first threat to be identified was the supersonic transport (SST), an aircraft that flies through the stratosphere at faster than the speed of sound. The exhausts from these aircraft are a source of nitrogen oxides (NO_x) which can chew up ozone in much the same way that chlorine atoms do. Then, nuclear weapons and the space shuttle were suggested as potential threats to the ozone layer, after which came the spray cans and even more recently, nitrogen fertilizers.

The scientific attempt to understand just what these disparate technologies were doing to the earth's atmosphere began in earnest in 1971 and it is not finished yet. The complexity of the chemistry going on in the atmosphere ensured that there would be pitfalls and surprises along the way. One of the most dramatic of these, which will be dealt with in more detail later, is worth mentioning here. As of this writing, it appears that the SST is much less of a problem than originally predicted. Continuing research resulted in a steady downward revision of the SST's predicted impact (an unusual and unexpected situation) until finally, data gathered in the last half of 1977 indicated that the SST might actually *produce* a small amount of ozone in the stratosphere instead of destroying it—this, despite an earlier three-year multimillion-dollar study by scientists around the world that concluded the SST was indeed a threat to ozone. Interestingly, the same research that resulted in a downward revision in the case of SSTs led to a substantial upward revision in the predicted impact of fluorocarbons.

It is, of course, the spray-can threat that has attracted most of the public attention. Indeed, the Spray-can War was characterized by a public interest in science that quite startled many of the researchers working on the problem. These scientists, accustomed to being treated with, at best, indifference and often with a veiled kind of hostility ("I never was much good at science"), found themselves actively, even eagerly, questioned by concerned and interested non-scientists. Perhaps it was because many people felt that, while decisions about SSTs and nuclear bombs were out of their hands, decisions about spray cans were not. And they were right—as the dwindling sales of aerosols and the booming business in roll-ons and pump sprays soon proved.

If the impact on the public was one notable feature of the ozone controversy, its impact on scientists was even more intriguing. A wide variety of researchers were attracted to the problem for a number of reasons. Many of them were truly worried about the possibly devastating consequences of ozone depletion. Moreover, the science involved was genuinely challenging: This was not a second-rate research problem. Yet it was also, at a time when society demanded relevance in research, unquestionably relevant. The solutions being sought were important to society and, perhaps more to the point, could clearly be seen to be so by the taxpayer. This had the happy consequence of loosening the purse strings and, with preferential funding, the ozone question acquired the distinction of being "where the action is." But there was also the glare of publicity and the strain of doing science on demand—or what one pressured participant angrily referred to as "science in a goldfish bowl." Nothing in their training had prepared them for this; in fact, they were predisposed to avoid it.

Nevertheless, many scientists expended a considerable amount of time and effort on the ozone problem—and not just on doing research. They testified before state and federal authorities, they served on scientific committees investigating the problem, they dealt with the press. At times, these efforts resembled nothing so much as a traveling road show and, for some, they became very nearly all-consuming. Nor was the pursuit of scientific truth in the matter of human technology vs. the ozone layer conducted without emotion and some acrimony.

This is partly because the scientific process is inherently one of confrontation. New suggestions or theories are often strongly challenged, and success goes to those whose ideas can withstand such attacks. In particular, success goes to those who have the best ideas first. The competition for scientific priority is always present, but is especially intense when the subject in question commands widespread interest and involves far-reaching consequences.

There are scientists who believe—or say they believe—that such confrontations and competitions are strictly impersonal. But this is clearly not true, certainly not in the case of the ozone debate, which was dominated throughout by strong personalities with strongly held convictions. One of them once remarked: "There are some real villains in this thing." It was a sentiment echoed by many others. The irony is that few people agreed on who were the villains and who the heroes.

Some scientists clearly prefer that the story of the ozone controversy be told without reference to these confrontations; they seem to shrink at the idea of allowing the public to see this human side of their work. But the story cannot be told this way, for it would not be the truth. These disputes were an integral part of the ozone story, as much a part of it as the balloon flights, the laboratory tests, the computer simulations, the scientific meetings, and the research papers that made it the "happening" it was.

Of course, scientists were not the only players in the game. One cannot ignore the role played by the aviation and fluorocarbon industries, both of which waged fierce fights to save their respective billion-dollar enterprises, nor that of the government agencies, the Congress, the state legislatures, and even some municipal governments. They all got into the act, partly because it was their ultimate responsibility to decide what must be done, and partly because, for them too, this was where the action and the money and the jurisdictional power resided. Few issues have so completely bridged the gaps among the "three solitudes"—industry, government and basic science—as the ozone controversy did.

The ozone war involved public relations as much as it did science; emotion as much as it did logic. Scientists would be accused of alarmism and publicity-seeking; industry of self-interest and deviousness; environmental groups and their "fellow travelers" of overreaction and hysteria; politicians and government agencies of inaction or self-serving manipulation.

Yet despite all the fireworks, the fundamental issue was a vital one: We simply had to understand the earth's atmosphere, to chart both its resiliency and its fragility, and to come to terms with our own, largely unwitting, ability to damage it. This was a formidable task, and the ozone scare forced us to make a start on it, but the job is far from done. Large global systems like the earth's atmosphere are characterized by exceedingly complex chemical and physical processes that frequently defy an adequate understanding, much less control. Chemicals like the spray-can propellants never existed in nature. In his increasing use of such chemicals, man seems at times to be engaged in some bizarre game of environmental Russian roulette.

It may be that, on a global scale, it is a game in which he is out of his league.

CHAPTER ONE

The Spray-can War Begins

The known fact is that fluorocarbon propellants, primarily used to dispense cosmetics, are breaking down the ozone layer. Without remedy, the result could be profound. . . . It's a simple case of negligible benefit measured against possible catastrophic risk both for individual citizens and for society. Our course of action seems clear beyond doubt.

> —ALEXANDER SCHMIDT, Food and Drug Administration, announcing a phase-out of fluorocarbon spray cans, October 1976.

In a way, Mrs. Lovelock started it all. Back in 1970, when her husband, Jim, decided he wanted to measure fluorocarbons in the earth's atmosphere, no one was much interested—certainly not the people who supplied funds for scientific research in his native England. So Mrs. Lovelock, the family business manager, broke out the grocery money, and her husband built a very sensitive instrument that soon detected minute amounts of fluorocarbons in the atmosphere. These chemicals did not come from nature; they were man-made, and it was not hard to figure out where they did come from. What Jim Lovelock was measuring was largely the accumulation of several decades' worth of hair spray and deodorant propellants, with perhaps a very small amount of refrigeration and air-conditioning coolants thrown in. Though neither he nor his wife could know it at the time, their modest investment in pure scientific research would threaten the billion-dollar refrigeration and spray-can industry and touch off one of the major environmental rows of the decade.

But Lovelock was no environmental crusader. He is an unassuming Englishman with modishly long graying hair and a soft, almost hushed voice. There is a gentleness about him that provides a striking contrast to the rather high-powered American scientists who dominate the ozone controversy. Lovelock is something of an oddity—a freelance scientist who works from a garage-*cum*-laboratory at his home in the rural English village of Bowerchalke. He prefers the precarious life of the free-lancer; security, he feels, kills scientific creativity and this, paradoxically, makes him profoundly nervous. With a Ph.D. in medicine he had been for twenty years a researcher at the British National Institute of Medical Research. It was a tenured civil-service position with a pension, safe and secure, but he could see it winding down inexorably to retirement, and so he got out. His independent philosophy occasionally affronts the sensibilities of the scientific establishment. Until he was able to acquire a respectable address through an honorary professorship at the University of Reading, he had trouble publishing scientific papers sent from his home address. The British magazine *New Scientist* reported that an editor of one scientific journal once told Lovelock, when he asked why a paper of his had been rejected, that "they were always getting crank papers from people living in the country."

When Lovelock set out in search of fluorocarbons, he was not looking to create another environmental scare story. On the contrary, in his early papers, published before the fluorocarbon threat to the ozone layer was recognized, he took pains to dismiss any suggestion that fluorocarbons could harm the environment. Lovelock subscribes to the "pure science" school of thought, which holds that curiosity is the only acceptable motivation for anyone truly worthy of the title "scientist." Motivations other than this make a scientist what Lovelock disdainfully refers to as a "professional science operator." In his scientific papers, Lovelock does note that the fluorocarbons might serve a useful purpose, acting as tracers that would allow meteorologists to track global air motions and wind directions. But pressed for his own motivations, he shrugs his shoulders.

For reasons he does not—perhaps cannot—enumerate, he was simply interested in measuring the chemicals.

Initially, he had trouble convincing anyone that the job was worth doing. His first applications for research grants were turned down flat. There is still a tone of outrage in his voice when he recalls the comments of one reviewer—"probably a distinguished scientist"—who dismissed the grant application as the most frivolous he had seen in a long time. This scientist doubted that fluorocarbons could be measured at all even at much higher concentrations than those Lovelock was claiming to be able to measure, and concluded that even if they could be measured, the project was still a waste of time. Lovelock speculates that there may have been some considerable chagrin at the granting agencies when the fluorocarbon controversy hit its full stride later on.

The refusal of research funds didn't dissuade Lovelock; in fact, it just made him more stubbornly determined to carry out the measurements. So with his wife's blessings and the housekeeping money —for which she received a gracious acknowledgment in one of his scientific papers—Lovelock built his fluorocarbon-measuring instrument. It incorporated an electron capture detector, a device invented previously by Lovelock that is among the most sensitive instruments ever designed for chemical analysis. It improved the ability to measure certain chemicals in the atmosphere by about one million times. Indeed, it was only because the instrument was so sensitive that Lovelock was able to measure the fluorocarbons at all.

Lovelock set up his first fluorocarbon monitoring station at his home. Before long, the family was forced to give up spray cans, not for environmental reasons but because they were interfering with Lovelock's backyard measurements. In 1971 and 1972, Lovelock made further measurements aboard a ship that sailed from Britain to the Antarctic and back, and in January 1973 he published the results in the British scientific journal *Nature*.

He had found only very tiny amounts of fluorocarbon in the atmosphere, and no one got very excited about it. In his *Nature* paper, Lovelock once again pointed out the potential usefulness of fluorocarbons as meteorological tracers, and once again he returned to his familiar theme that "the presence of these compounds constitutes no conceivable hazard."[1]

The statement could not, of course, have been more ill-advised, and it has many times returned to haunt him. "I boobed," he admits frankly. "It turned out I was sitting on a real bomb." But at the time, he had a very strong philosophical motive for writing the statement. "I'm not a doomwatch sort of person and I was most anxious that a straightforward investigation should not be turned into a doomwatch scare. I could easily have said these things may be a hazard to the health of future generations and this would have probably got me bags of support. I was sort of falling over backward. . . ."

With its implicit challenge to other scientists, Lovelock's statement virtually guaranteed that the fluorocarbon/ozone theory would eventually be discovered by someone. It was certainly what provoked Charles Kolb to start thinking about fluorocarbons late in the summer of 1973. Kolb, a physical chemist with Aerodyne Research Inc., a Boston contract research organization, felt the fluorocarbons had a "large potential for mischief," but he did not immediately perceive the threat to the ozone layer. Though he knew that the chemicals would release chlorine atoms, "at the time I didn't know about the efficiency of chlorine as an ozone-eater." So he did not really start to push his idea until late September, when he heard from Harvard scientist Mike McElroy about a meeting in Kyoto, Japan, earlier that month at which the problem of chlorine

[1] Interestingly, however, a group of meteorologists had already suggested fluorocarbons might be an environmental problem. The suggestion is in a little-known report on a meeting sponsored by the National Aeronautics and Space Administration in August 1971—before even the SST threat was identified. The report suggests that fluorocarbons would be broken up by ultraviolet radiation high in the atmosphere and that there didn't seem to be anything preventing them from getting up there. It also suggested that "an apparently irreversible alteration of the atmospheric composition" might be possible, but does not specifically mention ozone destruction, nor does it identify the mechanisms leading to this destruction. The fact that this idea originated in a workshop on how to measure pollution in the lower atmosphere probably accounts for the fact that it was not picked up by atmospheric chemists concerned with the upper atmosphere; as we shall see later, these two groups tended to be scientifically isolated from each other, particularly at the beginning of the ozone controversy.

in the earth's atmosphere had been discussed by scientists for the first time. However, none of the scientists present at that meeting had identified fluorocarbons as a potential source of chlorine that could destroy ozone. After Kolb learned of chlorine's special talent for destroying ozone, his interest in studying the fluorocarbons took on a greater urgency. In October he submitted a research proposal to the National Aeronautics and Space Administration, but the project was never funded. By that time, Sherry Rowland and Mario Molina, two chemists at the University of California's Irvine campus, were already hot on the trail.

For Sherry Rowland the story really started as far back as 1970. It started imperceptibly and, at least in the beginning, it unfolded slowly. If Rowland hadn't taken a particular train to Vienna; if he had not heard the early rumors about Lovelock's work; if he had not been searching around for something completely different to do; if he had not just taken on a bright young postdoctoral research associate who similarly was looking for something new—so much of what happened might not have, just as easily.

Rowland is formally known as F. S. Rowland or sometimes F. Sherwood Rowland. He prefers Sherry, which mystifies some of his colleagues. When one newspaper story referred to him as Mr. Rowland instead of Dr. Rowland, he wryly commented that he was "not concerned about the use of titles . . . except the occasional Ms. that I receive because of mistaken inferences about the gender of Sherry."

Rowland is a large man, more than six feet tall, with long, graying sideburns and a frequently furrowed brow. He carries his calculator in a hand-tooled leather holster slung around his waist. He wears size 14 shoes. The latter is not normally a statistic that matters to anyone, but Sherry Rowland's feet have acquired a certain notoriety in some scientific circles as a result of an escapade in Russia in 1967. Rowland was among a group of Western scientists to visit the Soviet science city of Novosibirsk that year, and, during their stay, they were challenged to a game of basketball by the Soviet graduate students. Rowland, who had played college basketball, was captain of the Western team.

The Soviets were able to scrounge gym clothes and running shoes for most, but they were not equal to the challenge posed by the size of Rowland's feet. So Rowland trotted out onto the floor barefoot—whereupon the Soviet captain immediately removed his own shoes.

After the game, Rowland returned to the locker room and calmly proceeded to peel a thick, single layer of skin off the sole of one foot. Unperturbed, he slapped the skin back on with cold cream and walked several miles in Moscow the next day. This sort of thing happened to him from time to time; his wife swears that before they were married, she had received letters from him written on the soles of his feet, or portions thereof.

Rowland was, in many ways, a most unlikely candidate as the originator of the fluorocarbon/ozone theory. Worrying about the ozone layer was a going concern long before he got into the game and, like Jim Lovelock, he was not a member of the clique of researchers who had staked out stratospheric chemistry as their special domain. This is one of the most striking features of the ozone controversy—the extent to which "outsiders" played a crucial role in identifying the threats to the ozone layer. Rowland was not an atmospheric scientist. He had specialized in the chemistry of radioactive isotopes and thus, in 1970, found himself in Salzburg, Austria, at an International Atomic Energy Agency meeting on the applications of radioactivity to the environment. He was feeling the need to renew himself scientifically and was at the Salzburg meeting on a fishing expedition for new ideas.

After the meeting, on his way to Vienna, he happened to share a train compartment with William Marlow of the U. S. Atomic Energy Commission (AEC). Marlow was responsible for organizing scientific meetings that brought meteorologists and chemists together, a task that rivaled in difficulty the mixing of oil and water. The ozone issue involved complex problems of both meteorology and chemistry, and the inability of these two groups of scientists to communicate with each other was to become one of the recurrent frustrations in dealing with the problem, and, as we shall see, a source of some spectacular personality conflicts.

Rowland expressed an interest in what Marlow was doing and, more than a year later, in January of 1972, attended one of Marlow's meetings in Fort Lauderdale, Florida. The AEC had been doing studies of the atmosphere for some time, although Rowland, whose work had long been supported by the AEC, was not involved. But because he had expressed some interest to Marlow, he was invited to the meeting.

There are two sources of new information at scientific meetings: the formal presentations and the scuttlebutt. It was the coffee-break chatter that provided Rowland with the information that Lovelock had measured fluorocarbons in the lower atmosphere in both the Northern and the Southern Hemispheres (data that would not be published in the scientific literature until a year later). Lester Machta, a scientist with the National Oceanic and Atmospheric Administration, remarked that the amount Lovelock had measured seemed to be about equal to the total amount of fluorocarbons that had been released into the atmosphere. Among the many ironies of the fluorocarbon debate was the fact that the source of this information was Ray McCarthy of E. I. Du Pont de Nemours and Company, the major producer of fluorocarbons, which they manufactured under the trade name Freon. McCarthy was lab director of Du Pont's Freon Products Division. As he remembers it, he, Lovelock, and Machta, who had been attending a conference together, went off hiking one day and, in a conversation later, began to talk about Lovelock's measurements. Lovelock asked McCarthy how much of the fluorocarbons had been produced and released worldwide, and McCarthy did a rough calculation.

It turned out that what Lovelock had measured seemed to be very close to the total amount that had been produced. The implication of this was that nothing was destroying the fluorocarbons— that they were just floating around in the atmosphere. But the chemist in Sherry Rowland would not let it go at that. The fluorocarbon molecules had to go somewhere. Maybe up? Rowland knew that "if you get high enough, there is radiation that will break up any molecule, so you know the molecules will come apart someplace. I can remember saying, of course it will always decompose with ultraviolet." (Ultraviolet radiation is at the high-energy end of the sun's spectrum.) So the seed had been planted. But it certainly did not fall into the "Eureka!" category. Rowland describes it as no more than a "casual thought" and certainly did not deduce from it that fluorocarbons would be a serious environmental threat.

"I had no thought at that point that it was anything except another molecule that had been put into the atmosphere and in obviously very, very low quantities. It wasn't until much later that one has any real feeling that there is anything serious about it."

In fact, he put the whole matter on the "back burner."

It was not until the summer of 1973 that it was revived, when Rowland was going through the annual ritual of preparing his research proposals for the following year. Rowland had been funded since 1956 by the Atomic Energy Commission (now the Energy Research and Development Administration). The AEC was supporting his research on the chemistry of atoms produced in nuclear reactions. However, over the years, Rowland had introduced new subjects into his research, including photochemistry (the action of light on chemicals) and the chemistry of fluorine, both of which laid the foundations for the fluorocarbon work that would follow. In the summer of 1973, Rowland called his sponsors at the AEC and told them he wanted to branch out. They seemed agreeable to the idea. He then wrote a proposal asking for money to study the fluorocarbons—a curious request, in a way, since it really had nothing even remotely to do with nuclear energy; but researchers who have proven to be productive in the field of basic research are often given a relatively free hand, since the benefits of such research are frequently unexpected and can neither be predicted nor planned. The AEC denied Rowland additional funds, but told him that if he wanted to use part of his regular funds for the study, they had no objections.

It is difficult to trace the reasoning that led from Rowland's "casual thought" in January 1972 to an actual research proposal in 1973. Since scientists were interested in using fluorocarbons as atmospheric tracers, Rowland thought that there might be some interesting chemistry in predicting what was going to happen to them. But he is emphatic that there were "no cosmic implications" behind the research proposal to AEC. Though he would later be teased by some colleagues about a subconscious recognition of a hot political issue, Rowland denies trying to jump on any environmental bandwagon. While his proposal to the AEC did suggest that the fluorocarbons would likely get into the upper atmosphere and be attacked by ultraviolet radiation, there was no suggestion that this would pose an environmental problem involving ozone.

Rowland was not unaware of the ozone issue. He had twice invited Harold Johnston, a chemist at the University of California's Berkeley campus, to come to the Irvine campus to discuss the supersonic transport (SST) controversy. This was a fight in which Johnston had been deeply involved for several years; he and many other scientists were engaged in a major $21 million program to study SST effects and were becoming increasingly convinced that large fleets of the planes would indeed pose a serious threat to the ozone layer. But when Rowland first set off on the trail of the fluorocarbons, he had no sense of being about to embroil himself in the ozone war.

On October 1, 1973, Mario Molina joined Rowland's team as a postdoctoral research associate. Molina is a native-born Mexican, the son of Mexico's ambassador to the Philippines. He attended boarding school in Switzerland and universities in Mexico, Paris, and Germany before entering a graduate program at the University of California's Berkeley campus in 1968. He received his Ph.D. early in 1973 for work on chemical lasers. Of the various research projects he talked over with Rowland, the fluorocarbon problem intrigued Molina the most. Atmospheric chemistry was entirely new to him—and it was a bit unusual for a young researcher and his senior colleague to embark on a project in which they were both essentially neophytes—but it seemed exactly the change of pace they were looking for. For Molina, too, the problem was "simply scientific curiosity. We took it as a challenge to chemists to see what happens to these things. We didn't even think about the environmental effects."

Rowland and Molina asked themselves a seemingly simple question: What happens to fluorocarbons once they are released into the earth's atmosphere? Rowland recalls that once they started working on the problem, "it unwrapped very quickly." The first thing they did was to look for "sinks" or removal processes that would destroy the chemicals in the lower part of the atmosphere. From the start, this seemed improbable. For one thing, fluorocarbons are inert—that is, they don't react chemically with anything.

It was their great advantage as propellants, for example, that they could be counted upon not to react with the products being propelled or with the things the products were being propelled at (including humans). But this inertness also eliminated many of the processes that could put the chemicals out of action in the lower atmosphere. Second, there were Lovelock's measurements, which seemed to suggest that virtually all the fluorocarbons that had thus far been released remained in the earth's atmosphere. If this was in fact true—it would later become one of the most contentious points of the controversy—it strongly argued against the existence of removal processes in the lower atmosphere. But Molina set out in search of them anyway.

He tried to think of all the things that could possibly destroy fluorocarbons in the lower atmosphere. The list of possibilities got shorter and shorter without any obvious fluorocarbon "sinks" turning up. It was a systematic approach to the problem and, Molina rather sheepishly admits, "a bit boring. It was rather frustrating because I kept coming up with all sorts of possibilities and working out each of them and saying no, it cannot be very important." The chemicals were not washed out of the troposphere by rain. They were not dissolved in the oceans. They were not removed by interacting with living things. In short, there appeared to be only one fate for the fluorocarbons—they would migrate upward. By November, Rowland and Molina had concluded that fluorocarbons would certainly reach the stratosphere. From about 15 to 20 miles up, no longer well shielded themselves by the ozone layer, the fluorocarbon molecules would be split apart by the sun's ultraviolet radiation. The whole process would, however, be an incredibly slow one; the calculations showed that the fluorocarbon molecules would stay around for 40 to 150 years before getting into the stratosphere and encountering the ultraviolet radiation that would break them up. Rowland knew that his decomposition of fluorocarbons would produce chlorine atoms.

At this point the two scientists debated writing up their research results. They had answered their original question: What happened to fluorocarbons? The chemicals went into the stratosphere and broke down to produce chlorine atoms. But there was a loose end: What happened to the chlorine atoms? Still no "cosmic implications." If Rowland had been asked at that point whether chlorine atoms would attack ozone, he would probably have said yes. But he would not have thought it that important, since there was so little fluorocarbon up there. A tiny amount of fluorocarbon would chew up only a tiny amount of ozone—or so he thought. Still, in the interests of thoroughness, Molina went off to find out what happened to the chlorine atoms. He sat down with pen and paper and worked out the way in which the various chemicals that are in the stratosphere might react with chlorine atoms. The next day, he came back and informed Rowland that there was a chain reaction involving chlorine. The chlorine atom would react with ozone, producing an ordinary oxygen molecule and a chlorine compound. But in another chemical step, the chlorine compound would be converted back to a chlorine atom. Each time this happened, an ozone molecule disappeared and the chlorine atom emerged intact. It would go on to destroy many thousands of ozone molecules before some other process managed to remove it from the stratosphere.

In working out the chlorine chain, Rowland and Molina were not being totally original; the fact that chlorine chain reactions could destroy ozone was already known to other scientists. In fact, Rowland and Molina were soon to find out that the importance of the chain in the earth's stratosphere was at that very moment a subject of debate within the scientific community concerned with the ozone layer (though, ironically, none of these experts had perceived the fluorocarbon threat). But Rowland and Molina did not belong to that community—their professional orbits had never overlapped those of atmospheric scientists—and they knew nothing of the chlorine debate or the scientists who were party to it.

The existence of the chlorine chain was a startling development but still not too worrying to Rowland and Molina. The fact remained that there just weren't a lot of fluorocarbons in the stratosphere. But would it stay that way? In short order, they had concluded that the answer to that question was probably no. Molina only began to get excited when he looked at the fluorocarbon industry's figures on annual production of the chemicals. The calculations indicated that a constant injection of fluorocarbons at the 1972 rates—about .8 million tons a year—would ultimately produce concentrations of fluorocarbons in the lower atmosphere 10 to 30 times the existing levels, amounting eventually to nearly 100 million tons. This in turn would produce prodigious amounts of chlorine in the stratosphere—something like .5 million tons. Their first rough estimates led them to conclude that as much as 20 to 40 per cent of the ozone layer might be destroyed. Moreover, it appeared that the destruction by chlorine would be at least comparable to the natural destruction of ozone and might well become the overriding influence. It might, in short, take over from nature.

They both had the same reaction to this: They had clearly made a big mistake somewhere. Molina thought it couldn't really be the problem it seemed to be because surely someone would have thought of it already. They each went off by themselves to think it over. They meticulously checked everything, did it over and over again. But they could find no flaw, and that's when the penny finally dropped. If this was really all there was to it, they'd found a serious problem.

Alarmed, they sought advice from the people who had been working directly on the ozone problem. Rowland called Hal Johnston and told him: "We've found a chlorine chain and a source of chlorine."

Johnston asked if Rowland knew that the chlorine chain was already known, that in fact there was quite a bit of discussion going on about it among the people working on stratospheric chemistry. "No," Rowland responded, "that's why we called you. We're trying to find out how much of this is known within the atmospheric-science community."

In the week between Christmas and New Year's of 1973, Rowland visited Johnston at Berkeley to find out what was being done by other scientists. Though many researchers had known for years that the chlorine chain would destroy ozone, no one had worried too much about it because there were no known sources of chlorine in the stratosphere. It was not until September of 1973, at the meeting held in Kyoto, that scientists publicly began to argue that chlorine chemistry might have some relevance to the real world, that it might actually be going on in the upper atmosphere and that this could spell trouble for the ozone layer. Johnston gave Rowland preprint copies of two soon-to-be-published scientific papers on the subject, one by a group of researchers at the University of Michigan and the other by a team at Harvard University. But these scientists were not talking about fluorocarbons as a source of chlorine in the stratosphere. They were talking about the space shuttle, a reusable space vehicle being built by the National Aeronautics and Space Administration, scheduled to go into operation in the 1980s. The shuttle's exhaust will produce hydrogen chloride, a source of chlorine atoms.

During their December meeting, Johnston also told Rowland about recently completed work by British researcher Michael Clyne that had the effect of making the chlorine problem even worse than Rowland and Molina had originally calculated. Johnston could not see anything wrong in what they had done, so Rowland left the meeting with a sense of urgency about publishing the data as soon as he could. He found himself in a paradoxical situation. He was about to take off to Vienna for six months on a Guggenheim fellowship, a trip that had originally been intended as a rejuvenating change of scene. He'd intended to use the opportunity to find a new scientific problem to work on that was different from what he'd been doing. The irony of the situation did not escape him; here he was, sitting on the newest and biggest problem he'd yet encountered and one, moreover, that gave every evidence of being a major bombshell. But he went to Vienna anyway and spent his first Sunday there, in the place where it had all started for him, writing the first paper on the fluorocarbon/ozone theory, which he promptly sent off to the British journal *Nature*.

In the paper, Rowland and Molina argued that virtually all the fluorocarbons that had ever been released were still in the lower atmosphere. All of it would eventually—and inevitably—reach the stratosphere, but this would happen slowly. Thus the full impact on the ozone layer would not be felt for perhaps several decades. In the emotionally neutral language of science, the authors also mildly observe that "if any chlorine atom effect on atmospheric ozone concentrations were to be observed from this source, the effect could be expected to intensify for some time thereafter." This was a significant point. The import of what they were saying was that the worst was yet to come and that there was not a thing anyone could do about it. Even if the fluorocarbons were banned immediately, the damage to the ozone layer would get worse before it got better because most of the chemicals already in the atmosphere hadn't even begun to do their dirty work.

Other researchers concurred. In September, the University of Michigan researchers published a paper in the American journal *Science* stating that if ground-level emissions of fluorocarbons were halted immediately, the destruction of ozone would peak in 1990 and "would remain significant for several decades." Their calculations indicated that chlorine destruction of ozone could take over from all natural-destruction processes as early as the first half of the 1980s.

Subsequently, the Harvard team published computer calculations that had produced several rather alarming scenarios. If releases of fluorocarbons continued to grow at their then current rate of about 10 per cent a year, the ozone layer would be depleted by 14 or 15 per cent by the year 2000. If fluorocarbons emissions growing at a 10 per cent annual rate were suddenly banned in 1978, ozone depletion would still grow to 3 per cent by the year 1990. And if the decision to eliminate the chemicals was delayed until 1995, "the reduction in ozone could exceed 10 per cent and would be significant for as long as 200 years."

The Molina/Rowland paper did not appear in *Nature* until late June, in part because the man who was handling it apparently disappeared with no forwarding address and the paper languished at the *Nature* offices for some time. Rowland was fretful about this at the time, but in retrospect he is thankful for the delay. It gave him and Molina much-needed time to gather their wits. They were not atmospheric scientists and, essentially interlopers, they felt they had a lot to learn before they crashed the stratospheric chemistry game.

In February they had a brief but enlightening taste of the controversy and even notoriety that was to come when a Swedish newspaper published a story about their theory. Rowland had been discussing the work with other scientists since Christmas, and the gossip mill had soon spread the news throughout the community. One of those to whom he sent a preprint of the *Nature* paper was Paul Crutzen, a Dutch-born meteorologist working in Sweden, who had identified the SST problem. Crutzen mentioned the Rowland/

Molina work in a speech at the Royal Swedish Academy of Sciences, and Katrin Hallman, an alert reporter for the newspaper *Svenska Dagbladet,* picked it up and made front-page headlines with it. (Crutzen had not known that there was a reporter in the audience, and he was a little chagrined at his role in helping Rowland and Molina scoop their own scientific paper.)

No one else in the media picked up the story at the time, but the incident taught Rowland a valuable lesson about the quick reflexes possessed by the fluorocarbon industry. He was still in Vienna at the time, and he heard about the Swedish story from a Du Pont public-relations man based in Geneva. The Du Pont Company is the largest single producer of fluorocarbons, and their Geneva man was rather anxious to discover if the Swedish story was true. Rowland said that indeed it was and proceeded to give the details. When he got off the phone, he went all over downtown Vienna searching for a copy of the Swedish paper and managed to find one for every day of that week except the right one.

Industry's grapevine was an efficient one. In March, during a scientific meeting at Berkeley, Molina happened to share a lunch table with two other people unknown to him. It was not long before he realized the two were from Du Pont—one was Raymond McCarthy—and they were discussing the news they'd recently received from Europe about the fluorocarbon/ozone theory. Molina got the impression that the whole thing had come as a complete shock to Du Pont. The shock was due not to the fact that industry had not considered the environmental impact of fluorocarbons, but to the fact that they had. In 1972, Du Pont had issued an invitation to fluorocarbon manufacturers around the world to attend a seminar on "the ecology of fluorocarbons." The invitations stated that the chemicals were being released into the atmosphere at a rate approaching a billion pounds a year, and they could be either accumulating in the atmosphere or returning to the earth's surface. "It is prudent that we investigate any effects which the compounds may produce on plants or animals now or in the future." The companies funded several research projects, and by 1974 the results of these studies indicated that fluorocarbons posed no major environmental problems in the lower atmosphere. The inert chemicals did not affect plants, played no part in the formation of smog, and were apparently not decomposed by chemical reactions near the earth's surface.[2]

Rowland and Molina, however, argued that this very inertness which seemed such a boon, simply transferred the environmental problem from the lower to the upper atmosphere and made it a global problem to boot. It was therefore hardly surprising that Molina sensed a certain gloomy frustration in the conversation going on next to him. But he did not break in, partly because he felt rather embarrassed at having inadvertently overheard it and partly because he and Rowland were still keeping a low profile. They were trying not to discuss the problem too much until they had some reassurances from the scientific community that they were not completely off the track.

The fluorocarbon story would ultimately generate several thick bindersful of press clippings, but it got off to a surprisingly slow start. Some articles appeared after the Rowland/Molina paper was published, but few of them got beyond California, and the story soon died. Several of the major newspapers, including the New York *Times* and the Washington *Post,* kept the story at arm's length. Later, the *Times'* science writer, Walter Sullivan explained why: There was so much "doomsday reporting" going on at the time that he was not particularly anxious to jump too quickly at this latest prediction of environmental disaster. This decision played an important role in the history of the fluorocarbon story; to the continuing frustration of many newspapers and science writers across the country, a science story often just doesn't become news until Walter Sullivan and the New York *Times* take notice of it.

The fluorocarbon story revived in the fall of 1974 when Dorothy Smith took an interest in it. Smith, the peppery, no-nonsense news manager for the American Chemical Society, is responsible for handling media coverage of ACS meetings. These meetings, which are held frequently and are often very large, can be overwhelming to cover, and Smith does an excellent job of alerting science writers to the highlights and organizing press conferences on controversial and interesting new subjects.

Rowland was scheduled to give a talk on the fluorocarbons at an

[2] Ironically, this was essentially the same process that Rowland and Molina went through.

ACS meeting in Atlantic City in September of 1974, and the abstract of his paper caught Smith's attention. She flagged it as one of the top twelve news stories of the conference and sent out an advance notice to the press. Her use of the word "Freon" in the release elicited a prompt call of protest from a spokesman for the fluorocarbon industry. Industry was becoming increasingly insistent that brand names for fluorocarbons not be used in discussions of the Rowland/Molina theory. That kind of publicity they could do without.

As Smith started arranging a press conference on the fluorocarbon problem for the approaching ACS meeting in Atlantic City, she received a call from the public-relations department of one of the major fluorocarbon companies. "I wouldn't call it pressuring," she recalled later. "He was leaning a little bit, but it was lightly done." The caller didn't see much reason for holding a press conference. He told Smith that there was no proof that the Rowland/Molina theory was correct, that no fluorocarbons had ever been detected in the stratosphere, and that there was no known mechanism to get them up there.

Smith thought this over and decided to check for herself. Normally she does not investigate the scientists she chooses for press conferences that closely—Rowland was, after all, an established researcher at a major university and had been active in the ACS— but she could see that this issue was already a touchy one, and she felt obligated to reassure herself. She called several scientists, and they all assured her that Rowland and Molina's work was not nonsense. One of them told her that Jim Lovelock had actually made one measurement of fluorocarbons in the lower stratosphere, although the data had not yet been published. None of them warned her off holding the press conference, so she decided to go ahead.

The ACS meeting was covered by the wire services, whose dispatches received relatively widespread distribution. But the story did not really take off until September 26, when the New York Times ran a front-page article by Walter Sullivan. The story was dominated by a discussion of the computer calculations done by the Harvard group headed by Mike McElroy. Rowland and Molina were barely mentioned in the story. Sullivan, who had not been at the ACS meeting, had not talked to them for the September 26 article, but he did call Rowland later that day for a follow-up story, which ran on September 27. Rowland can be philosophical about this. He realized that the press is constantly in search of the very latest information and the Harvard calculations represented the newest data around. (Rowland was fast becoming "media hip"—as *Rolling Stone* magazine would later refer to him.)

However, the *Times* story did exacerbate a growing feud between the Harvard and Michigan groups. We mentioned earlier the papers written by each of these two groups, giving their calculations of the amount of ozone depletion that might occur from fluorocarbon usage. Both papers were published in *Science:* the Michigan group's on September 27, 1974, and the Harvard group's on February 14, 1975.

At the time that Sullivan wrote his September 26 article focusing on the Harvard calculations, these data had not been published in the scientific literature. In fact, the Harvard paper was not even received by *Science* until three days after the *Times* story appeared, and it was not published in the journal, as we have noted, until February, some four months later. Thus the appearance of the information in the New York *Times* raised some eyebrows; the scientific community takes a dim view of the press usurping the role of the scientific journal.[3] But what really caused annoyance and bitterness was the fact that the *Times* story appeared just one day before the Michigan calculations were published in *Science,* in effect scooping them. (In fairness, it should be pointed out that the results of the Michigan study had also been published in the press. They were referred to in the Ann Arbor *News* about two weeks before they were published in *Science.* However, in this case, the Michigan paper had been accepted and scheduled for publication by *Science* long before; thus it was in the "preprint" stage. While not all scientists approve of press accounts of preprints, there seems to be less disapproval in this case than in the case of results that have not yet even been accepted by a scientific journal, primarily because the latter have not been officially reviewed by scientific peers.)

[3] The following note once appeared in *Physical Review Letters:* "As a matter of courtesy to fellow physicists it is customary for authors to see to it that releases to the public do not occur before the article appears in the scientific journal. Scientific discoveries are not the proper subject for newspaper scoops."

The *Times* article really signaled the beginning of the Spray-can War. Soon the story was picked up by Walter Cronkite of CBS-TV and by the newsmagazines, and there was no looking back.

The controversy profoundly disrupted the lives of the men who started it—Lovelock, Rowland, and Molina. Lovelock found the uproar an anathema. Like many other Britons, he thinks Americans have an unfortunate penchant for excess in environmental controversies, and he speaks with faint disapproval of the furor caused by the fluorocarbon issue in the United States. "The Americans tend to get in a wonderful state of panic over things like this," he once told a British newspaper.

"I respect Professor Rowland as a chemist, but I wish he wouldn't act like a missionary. . . . I think we need a bit of British caution on this."

He illustrated his point by referring to the controversy over mercury in fish. "The Americans banned tuna fish and they blamed industry until someone went to a museum and found a tuna fish from the last century with the same amount of methyl mercury in it."

Lovelock's choice of an example was ironic: He apparently did not realize that the "someone" to whom he referred was none other than Sherry Rowland himself. Two years earlier, Rowland and several colleagues published a paper in the journal *Science* that starts: "The mercury levels of museum specimens of seven tuna caught 62 to 93 years ago and a swordfish caught 25 years ago have been determined. . . . These levels are in the same range as those found in specimens caught recently." To further complete the irony, Rowland had received a modest amount of publicity for this work at the time and was briefly adopted as a champion of sorts by the anti-environmental camp.

Sherry Rowland adapted quickly and well to the intrusion of the world at large into his life. A controversy like that over the spray cans usually earns for its perpetrator a high-profile life-style that demands as much time in the congressional hearing room and the television studio as in the research lab. Some scientists react badly, refusing to make the transition or resenting it when they're forced to, but Rowland seemed to enjoy it immensely.

To all outside appearances, he is almost invariably relaxed and cheerful. He listens patiently, speaks calmly, and does not get hysterical. If Central Casting were looking for someone to play the role that Sherry Rowland played, Sherry Rowland would not be a bad choice. But the ozone controversy could provoke anger and tension in him too, and a perhaps inevitable obsessiveness. At one scientific meeting, he carried a loose and cumbersome bundle of files with him everywhere, as though fearful he would be caught off guard and unprepared to defend his position.

Rowland and Molina are in many ways a study in contrasts. Rowland is large and sometimes rumpled-looking; Molina, a dark, bearded young man, is slight and dapper. Rowland is gregarious and outgoing; Molina is somewhat reserved. But Molina stands his ground at the scientific meetings and the press conferences, challenging his opponents, senior though they may be, with an edge of defiance in his voice.

Still, you could hardly blame him for feeling perhaps a bit queasy at the outset. His Ph.D. was barely a year old when he was tossed into this dogfight, and he and Rowland were working in a field that was new to both of them. Their work was being rigorously scrutinized by the best brains in the business, who were there to pull it apart if they could. That is the way science works; Molina knew this and accepted it, but still it scared him a little.

Nor was Rowland immune to moments of tension, to moments of feeling that his scientific neck was on the chopping block. He and Molina had done a remarkably thorough job—they would be credited for that by many experts—and Rowland felt reasonably confident right from the beginning that they would not run into major obstacles. But there were moments of uncertainty. Once, in an introspective mood, Rowland muttered almost to himself that if the theory were proved wrong, he at least hoped it would not be because they left out something simple or obvious.

If, at the outset of the fracas, they could not see clearly what was ahead for them, they had at least been warned that there were confrontations aplenty to come, that there were few cease-fires in the ozone war. The warning came from Hal Johnston, their counterpart in the SST/ozone controversy. During their conversation at Christmastime, Johnston turned to Rowland and asked: "Are you ready for the heat?"

CHAPTER TWO

The What *Layer?*

Johnston's comment to Rowland captured the essence of the ozone controversy. The fluorocarbon debate would not be, as the SST debate had not been, simply a cool and dispassionate academic exercise carried out within the confines of the laboratory and the scientific meeting room. The ozone controversy was also an exciting but frequently emotional scientific battle. There were many reasons why this was so—reasons having to do with personalities, with the way in which science is done, and with the changing role of scientists in the social and political arena. But perhaps the overriding reason for the intensity of the debate was the fact that damage to the ozone layer is potentially so disastrous. It was true that the understanding of the ozone layer was less than complete and that there were even greater uncertainties in predicting exactly what would happen if it were damaged, but what they *didn't* know about the consequences of ozone depletion worried scientists even more than what they did know.

Most of the earth's ozone is located in the stratosphere, scattered thinly throughout a band about 6 miles thick. The maximum concentration it ever reaches is a few molecules of ozone to 1 million molecules of air (and the air up there is less than 5 per cent as dense as that at ground level). If all this stratospheric ozone existed in a single layer at sea-level pressure, it would be no more than .3 centimeter thick. (Under the same circumstances, the earth's entire atmosphere would be 8.34 kilometers thick.) Yet this thin and scattered band of molecules is absolutely essential to the existence of life on the earth's surface because it strongly absorbs the ultraviolet radiation from the sun that is most damaging to living things (called biologically active ultraviolet or UV-B).

Early in the earth's history there was no ozone shield, and the planet's surface was bathed in harsh ultraviolet light. Surprisingly enough, this radiation originally served a useful purpose since it provided energy to start the biochemical processes that led to life. But living organisms could not tolerate the high levels of UV-B and so required protection, such as that provided by the surface waters of the planet. Since visible light penetrates water to a much greater extent than does UV-B, a class of organisms gradually evolved that was capable of using this visible light for photosynthesis. (Today, most life forms are either capable of photosynthesis or dependent on organisms that are.)

Through photosynthesis, organisms began to produce oxygen for the earth's atmosphere, which, in turn, led to the formation of ozone. Once the ozone layer was in place to protect them from UV-B, living organisms were able to emerge from the oceans and evolve on land. It was, as atmospheric scientist Tom Donahue has put it, "an interesting bootstrapping kind of operation. The absence of ozone allowed life to begin. Its presence allows it to continue."

Since living organisms on this planet evolved over several billion years in an environment largely shielded from ultraviolet radiation —since many species in fact appear to be living right at the edge of their tolerance for UV-B—scientists are concerned about the possible removal of our ozone protection, even in some small degree.

What we do know about the dangers of UV-B stems from our study of the effects of the very small amount of this radiation that the ozone layer does let through. The known biological effects of UV-B, which makes up only 1 per cent of the total solar radiation to reach the earth's surface, are almost all bad.[1] It causes sunburn and is extremely destructive to living cells. It is strongly, almost compellingly, implicated as the cause of several skin maladies and diseases associated with exposure to the sun, including skin cancer, the most common of all cancers. There is strong evidence that skin damage from the sun is cumulative and irreversible; increasingly, doctors are warning that deliberate sun exposure is just asking for

[1] The most important beneficial effect of UV-B is that it triggers the production in the skin of vitamin D, which prevents rickets. However, large amounts of UV-B are not needed, and vitamin D deficiencies can be corrected with diet supplements if necessary.

trouble later in life. That "healthy" tan may not be such a good idea.

The most common forms of skin cancer are two rarely fatal varieties, referred to as nonmelanoma. In the United States there are two hundred to three hundred cases per one hundred thousand population each year. Nonmelanoma is easily diagnosed and can be successfully treated, but it is frequently disfiguring, and the treatment can be expensive and painful. Scientists have been able to make a strong case linking nonmelanoma to UV-B. There are several lines of evidence, including the fact that nonmelanoma occurs almost exclusively on sun-exposed parts of the body and is found most frequently among outdoor workers. Moreover, the incidence increases steadily as you move from the poles to the equator, consistent with the increase in ultraviolet radiation that also occurs with decreasing latitude.

UV-B is also implicated as the cause of a second kind of skin cancer, with the sinister-sounding name of malignant melanoma. Melanoma is less common and more deadly than nonmelanoma. A cancer of the pigment cells, it is fatal in about one third of the cases, representing a threat to life about equal to that of breast cancer in women. It was responsible for about 1 per cent of all cancer deaths in 1970. The average annual incidence in the United States at the beginning of this decade was about 4.2 cases per 100,000 population.

The case linking melanoma to UV-B is neither as strong nor as clear-cut as that for nonmelanoma. Like nonmelanoma, it occurs frequently at low latitudes, and there is some evidence to suggest that sun-exposed parts of the body are the most susceptible. However, melanoma *can* occur on parts of the body normally protected by clothing, and the scientific evidence suggests that the dependence of melanoma on exposure to sunlight and UV-B is more complex than that of nonmelanoma.[2]

Another important factor linking both forms of skin cancer with ultraviolet radiation is the fact that the disease is rarely suffered by

[2] It appears that the incidence of melanoma has been increasing, possibly as a result of increased sun exposure due to life-style changes such as increasing outdoor recreation and changes in clothing styles. In its 1976 study, the National Academy of Sciences was sufficiently alarmed by this trend, quite aside from the ozone controversy, that it urged a public education program to alert people to the dangers of sun exposure and to the signs of melanoma.

dark-skinned people, whose pigmentation provides an effective ultraviolet screen. (A tan is a less effective temporary screen.) Both the dark-skinned tropical races and the light-skinned northern races are that way for very good adaptive reasons. In the tropics, the sun penetrates the smallest amount of ozone and the intensity of UV-B at ground level is high, so the people there are black. As you progress toward the poles, the sun's rays strike the earth at a greater angle, slanting through ever greater amounts of ozone. The UV-B at the ground becomes weaker, and the skin becomes paler. The fair-skinned Celts have evolved in a way that allows them to soak up all the limited ultraviolet radiation available to produce vitamin D. Transplanted South, the Celt is a bad sunburn risk and a prime candidate for skin cancer; those who have moved to Australia have provided scientists with the best skin-cancer statistics in the world. Conversely, blacks who move to high latitudes face potential problems of vitamin D deficiency, but this problem can be easily corrected through diet.

The conclusion to be drawn from the scientific evidence to date is that an increase in the incidence of both nonmelanoma and melanoma is a highly probable consequence of increased ultraviolet radiation. It should be noted, however, that this has not absolutely been proved, particularly for melanoma. UV-B levels at the earth's surface would increase 2 per cent for each per cent reduction in the ozone layer. Estimates indicate that a 10 per cent reduction in ozone (and therefore a 20 per cent increase in UV-B) would cause a 20 to 40 per cent rise in the incidence of skin cancer.

It is easy to see why the skin-cancer issue received so much attention during the ozone controversy. It most directly affects humans and is, despite the uncertainties, the best understood of all the effects a weakened ozone layer might be expected to produce. Moreover, the emotion attached to the word "cancer" proved to be a powerful tool in a controversy that was as much political as it was scientific. But there is a danger in human self-centeredness. Human beings are part of the global ecosystem not above it, and it is possible that the most devastating consequences of ozone depletion will hit us indirectly, through its effects on plants, animals, and the climate. Thus, while the emphasis on skin cancer was hardly surprising, it was in many ways rather unfortunate, for this may not be the most serious problem. Confronted with an increased level of

UV-B, human beings can, at least up to a point, do something about it. We can choose not to stay too long in the sun. We can wear protective clothing or develop effective sun-screening lotions.[3] Many of the other living things with which we share this planet, and on whom we depend, are not as mobile nor as readily protected. Plants and animals do have protective coverings, such as fur, feathers, or special pigments; the shine on an apple, for example, is produced by a substance that essentially acts as a sunburn lotion. But these protective coverings are adapted to the present levels of UV-B and may not be effective against increased amounts of radiation.

Deliberate avoidance of UV-B is not as easy as it might seem, even for mobile creatures. Few organisms sense ultraviolet radiation directly—they do not see or feel it as they do the sun's visible light or heat—so they would have no way of knowing that it has increased. Humans can build instruments to "see" UV-B, but most organisms take their cue from the sun's heat and light. These cues would not be altered by changes in the ozone layer, and so most organisms would have no motivation to avoid the attendant higher levels of UV-B.

There are other methods of coping with increased levels of UV-B. One is to repair the damage done to cells; the other is evolution. Evolution is clearly useless as a short-term strategy. It takes too much time; it took organisms millions of years to evolve to their present level of UV tolerance. Man may now be capable of upsetting the ozone balance over a period of decades or centuries, which is sudden indeed in evolutionary terms. Most organisms cannot hope to adapt that quickly. Nor can we.

Repair processes, on the other hand, may offer at least a limited defense. UV-B can break up the molecules in living tissues, including the DNA molecule responsible for passing genetic information from parent to offspring. In normal cells, the repair processes can undo more than 99 per cent of this damage. But the repair capability of many species seems to be taxed just about to the limit, and increases in UV-B could conceivably tip the scales.

[3] Whether humans *would* protect themselves remains an open question. Queensland, Australia, has the highest skin-cancer incidence in the world, but the people there do not protect themselves any more than less vulnerable populations.

Scientists cannot predict in detail what would happen to the global ecosystem if the ozone shield were to be weakened, but the evidence they have accumulated makes them fearful. Studies of plants, for example, indicate that increased levels of UV-B may cause cell mutations, affect growth and development, and reduce the capacity for photosynthesis, the process by which plants produce oxygen for the atmosphere and convert carbon dioxide into organic material—both essential for life on this planet.

Scientists were also very worried about what would happen to lower forms of life. They were particularly concerned about phytoplankton, the tiny plants in the oceans that form the foundation of the aquatic food chain and produce a significant amount of oxygen for the earth's atmosphere through photosynthesis. How they would survive increased levels of UV-B is unknown. Since water can also act as a UV shield, the plankton could theoretically avoid danger by migrating to greater depths in the ocean. But they are unlikely to do so, since it appears they cannot "see" UV-B. Even if they did move deeper in the water, they would face a harsher environment and a reduction in the amount of visible light available for photosynthesis. Either way, their vital contribution to the earth's ecosystem would be reduced.

Thus, though the catalogue of unknowns is admittedly large, all of the evidence indicates that the consequences of ozone depletion would be very grim. Indeed, there may be a lesson for us in earth's prehistory. For many years, scientists have tried to explain the massive extinction of species that occurred about 65 million years ago, at the end of the Cretaceous period. Fully one third of the species, notably the dinosaurs, suddenly and inexplicably died out. Scientists have never been able to come up with an entirely satisfactory cause-and-effect explanation for this. Some researchers have now suggested that ozone depletion may have been responsible. The depletion was due to natural causes, of course. The researchers suggest a scenario like this: Cosmic radiation is produced when the sun flares up or when a star explodes in a supernova. This radiation streams through space, eventually hitting the earth's upper atmosphere, where it produces large amounts of NO_x, which then leads to ozone destruction. Although this theory is still speculative, if it is true, it provides a striking example of the impact that NO_x can have on the ozone shield. The scientists who proposed this idea say in a scientific paper that the current concerns about man-made destruction of ozone "may be well founded, since it is possible that major ozone depletions occurring in the distant past have had a profound effect on the development of life as we know it."[4]

Biological problems are not the only ones associated with changes in the ozone layer. There is also the climate problem. Some scientists believe that the earth's climate is changing, perhaps drastically, at least in part because of human activities. There are many possible causes; ozone depletion would be but one, and predicting its contribution to the overall climatic problem is, if anything, even more difficult than predicting its biological consequences. Ozone controls the temperature of the stratosphere and therefore its climate. Though changes in the ozone layer would unquestionably alter the climate of the upper atmosphere, its effect on the climate and weather conditions in the lower atmosphere is largely unknown. Lack of knowledge does not, however, imply that the effect on climate will be nonexistent or even negligible. In general, scientists do not consider even "small" man-made alterations in the earth's climate to be prudent. Actually, the use of the word "small" in connection with climatic change can be deceptive. A change of one or two degrees might seem small—and it is if it occurs in a localized area over a period of a few days or a week. But a drop of a few degrees in the average annual temperature over the entire globe is a serious matter indeed; for example, there is a difference of only about five degrees between the earth's present temperature and that of full ice age conditions.

Even seemingly small climatic changes can, and probably will, produce a multitude of problems, but perhaps the most serious would be the impact on global food production. Agriculture is extremely sensitive to climatic variations, particularly in mid-to-high latitudes, where the variations tend to be more pronounced and where growing conditions are often marginal. The sensitivity of much of the world to climatic change is so great that we cannot afford to view any perturbation with equanimity.

If predicting the consequences of ozone depletion has proved to be a perplexing and uncertain business at best, it is misleading to

[4] The paper was published in *Nature* by George Reid and I. S. Isaksen of the National Oceanic and Atmospheric Administration and T. E. Holzer and Paul Crutzen of the National Center for Atmospheric Research.

suggest that complete ignorance characterizes the ozone problem. On the contrary, the ability of scientists even to perceive the threat to the ozone layer presupposes a substantial knowledge of the atmosphere and how it works. In fact, interest in the stratosphere and the ozone layer began at the end of the last century largely as a strictly academic exercise to satisfy the persistent curiosity of scientists concerning the earth's atmosphere. At that time, the prevailing opinion and all the evidence indicated that the temperature of the atmosphere decreased as you ascended—any mountain climber could attest to that. It would not have been illogical to assume that the temperature continued to drop with height to the reaches of outer space. But assumptions were not good enough for the early atmospheric scientists; they wanted to measure the temperature changes directly. They began strapping themselves into balloons so they could extend their measurements ever higher into the atmosphere. The history of this early ballooning is a fascinating tale in itself.

The first balloon flight made for scientific purposes occurred in 1784 from London. By the late 1800s, a large number of scientific flights were made, many of them carrying researchers to heights where they began to experience oxygen starvation. Indeed, several died for their trouble; others, more fortunate, simply passed out from lack of oxygen and recovered as their balloons slowly descended. Their measurements indicated that the temperature continued to drop with height.

Undaunted, other researchers persisted with remarkable tenacity in their attempts to reach higher altitudes. They developed automatic instruments that could take measurements without the need for human observers, but ran into one annoying technical snag—the ink in the pens of the measuring instruments often froze up.

A Frenchman named Teisserenc de Bort, one of the leading investigators of the time, used both balloons and kites for his observations. He had some problems with his kites.[5] Once several of them escaped and went floating over Paris, trailing miles of wire. This cut off communications between Paris and Rennes on the day that Paris was awaiting news of the outcome of the Dreyfus trial. De Bort had better luck with his balloons. By 1902, he had made more than two

[5] A description of the experiments of de Bort and others can be found in the book *The Edge of Space* by Richard A. Craig.

hundred successful flights, and in April of that year he reported to the French Academy of Sciences that the temperature of the atmosphere stopped decreasing about seven miles above the surface; in fact, the measurements indicated a slight increase above that height. Initially, his conclusions met with considerable skepticism, but soon other researchers confirmed that, above seven miles, the atmospheric temperature does indeed start to rise. Scientists had discovered the stratosphere.

They continued to extend their atmospheric observations with advanced balloons (and, after World War II, with rockets) and found that the temperature continues to rise with height and reached a maximum in the stratosphere at about thirty miles up, where it was almost as warm as the earth's surface.

The difference in the way the temperature changes with altitude led scientists to divide the atmosphere into distinct regions. The region between the earth's surface and the stratosphere, characterized by a drop in temperature with height, was called the troposphere. The differences between the behavior of temperature in the troposphere and in the stratosphere have profound consequences in terms of atmospheric pollution. In the troposphere, cold air is above warmer air. Since the cold air is denser, it tends to sink, while the warm air tries to rise. Thus there is a good deal of vertical mixing of the air, and the result is that pollutants are carried down to the ground and removed. Moreover, almost all precipitation occurs in the troposphere, and pollutants are rapidly rained out. The troposphere can cleanse itself of pollutants in about a week or so.

In the stratosphere, on the other hand, warm air rests on top of colder, denser air. This is a stable situation known as an inversion, and there is no tendency for the air to mix.[6] Also, the stratosphere is very dry and has no rains to wash out pollutants. As a result, pollutants that would take only a week to wash out of the troposphere will remain in the stratosphere for several years, and they therefore have about a hundred times as long to do their dirty work in the upper atmosphere. A National Academy of Sciences study noted that "the stratosphere can be likened to a city whose garbage is collected every few years instead of daily."

[6] This situation occasionally occurs in the troposphere and gives rise to serious urban pollution problems, such as those experienced by Los Angeles.

Ironically enough, it is the presence of ozone in the stratosphere that is largely responsible for this state of affairs. The temperature increase in this region is caused by the heat given off in the chemical reactions that form and destroy ozone. Thus the ozone layer is, in a sense, its own worst enemy, since it creates the temperature conditions that make the stratosphere such a good trap for ozone-destroying pollutants.

As mentioned earlier, the total amount of ozone in the stratosphere is very tiny—less than one ten-thousandth of 1 per cent of the air in that region—and it is the result of a balance between processes that form it and others that destroy it. The formation of ozone was first explained in 1930 by Sidney Chapman, an Englishman working at Oxford University, who has become known as the father of aeronomy (the science of the atmosphere). Chapman suggested that oxygen molecules would absorb part of the sun's ultraviolet radiation and split apart into two oxygen atoms.[7]

UV—light   oxygen molecule   oxygen atoms

These separated atoms would then combine with another oxygen molecule to form ozone, which contains a total of three oxygen atoms.

oxygen atom   oxygen molecule   ozone

Production of ozone thus requires two ingredients—ultraviolet radiation and supply of oxygen molecules. The intensity of the ultraviolet radiation increases the higher you go in the atmosphere, but

[7] Oxygen molecules absorb a different wavelength of ultraviolet than do the ozone molecules. Thus oxygen molecules do not protect life on the planet's surface from UV-B.

the supply of oxygen molecules steadily decreases with height. The product of the two ingredients hits a maximum in the stratosphere, and that is where most of the ozone is made. Since neither the strength of the sun's radiation nor the amount of normal oxygen in the atmosphere can be changed by human intervention, there is nothing we can do to alter the rate at which ozone is produced.

If nothing in nature destroyed ozone, all of the oxygen in the stratosphere would eventually be converted to ozone. But Chapman knew that oxygen atoms could destroy ozone as well as make it, by colliding with the ozone molecule and breaking it apart to form oxygen molecules once again.

oxygen atom   ozone   oxygen   oxygen
                                  molecule   molecule

The speed of this ozone-destroying process is directly proportional to the amount of ozone present; the more ozone there is, the faster it is destroyed. Therefore, conditions in the stratosphere eventually reach a point where ozone is being destroyed at the same rate as it is being made; this is called a "steady state," and it determines the total amount of ozone that will exist at any given time.

A steady state is something like the situation that would occur if you tried to fill a bucket that had a hole in the bottom. If you poured the water in at the top at a fixed rate, the water level in the bucket would start to rise. But the fuller it got, the faster the water would run out of the hole. Eventually the water would run out of the hole as fast as you were pouring it in, and a steady state would be reached in which the level of the water would neither go up nor down. If the hole were made larger, the level would start to drop. But as it dropped, the rate at which it ran out of the hole would slow down and eventually a new, but lower, steady level would be reached when the two rates again balanced each other.

Chapman was highly successful in explaining why ozone existed in the stratosphere and roughly how it was distributed, but scientists soon discovered that this was not the complete story. Measurements showed that the total amount of ozone was much lower than that predicted using Chapman's simple chemistry. Oxygen atoms were not destroying ozone as fast as it was being made; the destruction rate was in fact only 20 per cent of the formation rate. Clearly something else had to be chewing up ozone. It could not be the major components of air (oxygen and nitrogen molecules), because they do not react with ozone; nor were the minor components, such as water vapor and carbon dioxide, likely candidates, for the same reason. The only remaining possibility was that ozone was being destroyed by trace substances in the stratosphere. These substances are present in very tiny amounts—much smaller concentrations than even ozone itself—and if they were indeed the culprits, scientists were presented with a dilemma: How could something present in trace amounts destroy much larger amounts of ozone? The answer is found in what are known as catalytic chain reactions.

A catalytic chain is a series of two or more chemical reactions in which a substance (called a catalyst) destroys another substance (e.g., ozone) without itself being destroyed. The catalyst emerges from the reaction intact, ready and able to continue its rounds of destruction. As the process repeats over and over again, each molecule of the catalyst chews up literally thousands of ozone molecules. The chain is broken only when some other competing process manages to remove the catalyst (a situation that occurs only about once every few thousand times around the cycle).

The existence of catalytic chain reactions made nonsense of the frequently expressed belief that human activity could not possibly damage the earth's atmosphere because the atmosphere is such a large system. It meant that even comparatively small amounts of pollutants—amounts that man seems perfectly capable of injecting into the stratosphere—could have devastating global effects.

Laboratory studies showed that certain oxides of nitrogen (called NO_x) and certain oxides of hydrogen (called HO_x) were effective catalysts. In the 1950s, David Bates, an applied mathematician from Belfast, and Marcel Nicolet, an aeronomer from Brussels, recognized the importance of these chemical reactions and showed how NO_x and HO_x could be formed in the atmosphere in the region above the stratosphere. But they did not suggest how these substances might affect the ozone in the stratosphere.

John Hampson, an English-born scientist working at the Canadian Armaments Research and Development Establishment (CARDE) in Quebec, was the first to suggest how HO_x could be formed in the stratosphere. He did some calculations of the effect on ozone and suggested that HO_x might be the elusive catalyst primarily responsible for natural destruction of ozone in the stratosphere. Unfortunately, he published his results in a CARDE report that did not receive wide scientific recognition. However, the matter was pursued by B. G. Hunt, a scientist with the Australian Weapons Research Establishment on sabbatical leave in the United States, whose more detailed calculations of the HO_x/ozone effect were published in a scientific journal. (Hampson and Hunt were studying the upper atmosphere because their respective institutions were at that time interested in the problem of re-entry into the atmosphere of intercontinental ballistic missiles.)

HO_x did not, however, turn out to be the only catalyst for ozone destruction. In 1970 Paul Crutzen suggested that a catalytic chain involving NO_x could be important. Until fairly recently, it was thought that the NO_x and HO_x chains together could provide the balance between the rate at which ozone was formed and destroyed. But, as we have seen, chlorine atoms can also destroy ozone by a catalytic chain. Originally, it was thought that chlorine from natural sources was not making a major contribution to the ozone budget. But, by mid-1977, new laboratory measurements began to suggest that chlorine chains may be more important and NO_x chains less important than previously believed. Thus, it may be some time before the relative contributions of NO_x, HO_x, and ClO_x become clearly established.

The supersonic transport was the first example of human technology to be accused of harming the ozone layer. First, John Hampson suggested that water vapor (which give rise to HO_x) might pose a problem. Then, Paul Crutzen suggested that the amount of NO_x that the projected fleet of SSTs would emit was comparable to the amount naturally produced in the atmosphere and might therefore affect the ozone layer. But neither of these suggestions initially received much attention. In fact, the SST/ozone issue did not really become a major controversy until 1971—and then it happened because of a meeting in the men's washroom.

CHAPTER THREE

The Meeting in the Men's Washroom

I would be very much obliged if you refrained from referring to a matter with which you are not familiar. The arguments you put forward against supersonic aircraft are totally devoid of sense.

> —Letter from Pierre Cot, director-general of Air France,
> in response to a letter expressing concern
> about the Concorde's environmental impact.

The SST was in trouble in America. By early 1971, a diffuse but highly vocal environmental lobby was slowly gathering itself for a major assault. Their target was not solely—or even primarily—the Concorde, a *fait accompli*. The United States also had an SST in the works, and one that promised to be a more serious polluter than the Concorde. In March 1971, congressional hearings were held on a bill to continue funding The Boeing Company, which was building two SST prototypes. A crucial vote on the bill was due in the House of Representatives before the end of March.

One of the witnesses at the hearing was James McDonald, an atmospheric physicist from the University of Arizona. McDonald was passionately opposed to the SST because he had become convinced that supersonic operations could well lead to increases in the incidence of skin cancer and other damaging biological effects. He urged the congressmen to reject the bill.

McDonald came under sharp questioning, but the congressmen seemed more interested in his views on unidentified flying objects than they were in his concerns about SSTs. McDonald had, in fact, been interested in the UFO problem for some time. He had done a study of UFO data, believed the problem to have been "scientifically ignored," and had been a vocal opponent of plans to cancel a UFO observation program. But McDonald had come to the hearing to talk about SSTs and, as he continued to be questioned about his interest in "flying saucers," he became impatient and annoyed. "You use the term 'flying saucers.' You use the term 'believe.' I don't use those terms," he told his questioners. "It is not entirely clear that there is a relationship between SSTs and UFOs," he added with some exasperation, implying that the questions were irrelevant to the subject at hand, namely the fate of the U.S. SST program.

"I think there is a relationship," Congressman Silvio Conte of Massachusetts responded tartly. To Conte, it was a matter of self-protection. If he voted against the SST on the strength of McDonald's testimony, he didn't want to have to contend with having it thrown back at him on the floor of the House that McDonald was a UFO nut.

McDonald would run into many difficulties in consequence of his interest in UFOs, which had not then gained even the very limited vogue among scientists that it would later acquire. But he was not a flying-saucer nut; his scientific colleagues for the most part considered him a serious and conscientious scientist, and they knew he had not simply pulled the skin-cancer theory out of a hat.

McDonald had, in fact, been studying the environmental effects of SSTs since the mid-1960s at the request of the National Academy of Sciences. As a member of the Academy's panel on weather and climate modification, he had been largely responsible for debunking arguments that SST water-vapor emissions in the stratosphere would cause climatic changes. It had been suggested that the water vapor would produce contrails and persistent, hazy veils of slowly falling ice crystals, causing changes in the heat balance of the atmosphere. McDonald specialized in the properties of ice crystals, and it was his calculations that led the National Academy to conclude in 1966 that this SST environmental problem was not serious.

In the summer of 1970, the Academy had asked McDonald to re-examine the SST's stratospheric effects, and it was in the course of this work that McDonald began to make connections among SST flights, ozone reduction, increased ultraviolet radiation, and skin cancer. McDonald based his estimates on a projected 1985-90 fleet of eight hundred SSTs of the U.S. type (which would be larger than the French-British Concorde, would fly higher in the stratosphere and would pose a greater threat to the ozone layer). Economic projection being the inexact science it is, the future size of SST fleets was always a rather fluid figure but McDonald's figure was typical of projections at the time. In 1971, the Boeing SST project, which was running into technical and financial difficulties, was lagging behind the Concorde program and the Soviet TU-144 program.

McDonald estimated that eight hundred U.S.-type SSTs would cause a 4 per cent decrease in ozone, but he used a more conservative figure of 1 per cent to calculate an increase in skin cancer of five thousand to ten thousand new cases a year in the United States alone. (As it turned out, these estimates were not wildly out of line with subsequent Department of Transportation and National Academy of Sciences estimates of six thousand to eight thousand new cases a year for each per cent of ozone reduction.)

McDonald argued that these estimates, though tentative, were much too disturbing to warrant the continuing trend toward SST technology. Two prototypes, he acknowledged, posed no skin-cancer hazard, but he urged Congress to think twice about an irreversible commitment to the SST that would carry them beyond the point of no return.

McDonald considered skin cancer the "hidden SST problem." Other researchers had previously concluded that SST operations might have some effect on the ozone layer but largely dismissed the problem because these changes seemed small and insignificant. McDonald's argument was that even these seemingly small changes in the ozone layer could not be ignored because of the skin-cancer problem and because of the biological consequences of increased ultraviolet radiation for all living things.

McDonald in fact had done a sound analysis of the skin-cancer problem using perfectly acceptable scientific reasoning and the best scientific data available at the time. The evidence he used—the correlation between skin cancer and latitude, the greater number of cases among outdoor workers, the increased incidence on sun-exposed parts of the body, and the greater vulnerability of light-

skinned people—presaged the arguments and evidence that would
be employed in subsequent full-scale studies. Since McDonald did
his analysis, at least three major studies have examined the ozone/
skin-cancer problem, and their conclusions vindicate McDonald's
early concerns.

Ian Clark, a Canadian scientist who knew McDonald, says in an
unpublished manuscript about the SST controversy, called "Technical Advice and Government Action": "Contrary to the impression of many SST supporters, McDonald did not hastily throw together some speculative estimates of skin-cancer incidence and rush
to the press." He reported his concerns privately to the Department
of Transportation in November 1970, and also to the National
Academy. Clark notes that "although none of the substantive parts
of the analysis were refuted [by the Academy], McDonald was cautioned about its political sensitivity and admonished to subject the
analysis to the regular scientific review process before making the
conclusions public."

Thus it was not until the March 1971 congressional hearing that
McDonald did go public with the skin-cancer problem, and by that
time he was clearly very worked up. His testimony was lengthy, and
he soon had the congressmen awash in rather complex technical detail. He spoke so quickly they had to ask him to slow down. Apolo-
gizing, he said: "There is a lot to be said here. It hasn't been said to
any congressional audience, let alone the public, and this is the first
time these months of research have had any airing.

"The whole history of evolution . . . has been a battle with ultraviolet," McDonald said. "We've always just barely won. . . ."

A portrait of McDonald's testimony is revealed in notes taken by
another scientist who was present at the time. Will Kellogg, associate director of the National Center for Atmospheric Research, was
inclined to believe that the impact of SSTs would not be significant,
and he had previously stated that there were no environmental
reasons to delay the construction of two prototypes. He had also
strongly endorsed a statement that rejected the idea of canceling
the SST, favoring instead a research program to ensure that there
would be no undesirable environmental side effects. Kellogg was extremely skeptical about the skin-cancer theory and was rather put
off by the manner of McDonald's presentation. Yet Kellogg's reac-

tions demonstrate what is most admirable about the scientific mind
—an unwillingness to reject a new idea out of hand, however bizarre or improbable it might seem, and a willingness to put aside
emotion, gut feelings—even, perhaps, one's own biases—to grant a
new theory its right to rigorous scientific investigation. Kellogg
wrote of McDonald's presentation:

"Angry and finger-wagging, clouding the real issues by constantly
implying that others who have looked at it are either ignorant or
dishonest. He is a good scientist and most (though not all!) of his
facts are correct—but he presents the case with too much emotion
and a rather transparent hostility to the whole idea of the SST. In
other words, he is out 'to make a case against' instead of trying to
arrive at a fair assessment of the probable effects. I think that his
digression on the dangers inherent in new technology—a theme that
repeats itself throughout—makes it clear that he had a bias even before looking into the matter and so he was seeking arguments
against the SST from the start. While his statement is slanted, I do
think that each of the points he has raised deserve careful consideration by the scientists who can help to clarify the situation. Jim
McDonald would not have raised these worries without having
some good arguments to back him up."

During the course of his testimony, McDonald had been asked if
he thought his interest in UFOs had created a credibility gap for
him. "Not that I'm aware of," he responded blandly. Yet the New
York *Times* would later report, after the March hearings, that "the
weight of his findings was sharply discounted after he acknowledged, under questioning, that he had earlier testified before
Congress to his serious concern about unidentified flying objects and
the possibility that 'flying saucers' may have been related to power
failures in New York City."

During the March hearings, McDonald noted that he had presented his finding to the DOT the previous November but he apparently didn't have much success in impressing them with the
skin-cancer problem. He was asked if there had been any follow-up.
"None that I have heard of and certainly none that came to me."
What had DOT done about the matter? "Presumably not very
much." Subsequent to McDonald's presentation at DOT, Trans-

portation Secretary John Volpe, an ardent SST supporter, had said that the cancer problem could be dismissed "categorically," and the DOT issued an SST environmental impact statement that made no mention of skin cancer.

Nevertheless, DOT had, possibly unwittingly, set in motion a series of events that would bring the skin-cancer issue to prominence.

In September of 1970, the Department of Commerce Technical Advisory Board (CTAB), responding to a request from the Department of Transportation, set up a panel on SST environmental research. Its function was to determine whether there was enough scientific information to answer the questions being raised about the SST's environmental impact and whether research programs being planned by DOT could be expected to provide adequate answers. The emphasis was largely on the potential modification of weather and climate by SST operations.

CTAB's SST panel included a physicist, a physical chemist, a radiologist, a professor of forest ecology, a chemist, two meteorologists, and a vice president of Trans World Airlines (one of the first airlines to place an option for the U.S. SST, in 1963). No one on the panel had ever done any research on the impact of SSTs on the stratosphere, but this was not unusual for such an advisory panel. There is a school of thought in Washington that holds that it is best to fill such panels with scientists who are not personally involved in the relevant research, on the grounds that they have no axes to grind. The obvious drawback is that they also generally know very little about the subject at hand.

The CTAB panel held its first meeting late in 1970, with a second scheduled for January 23, 1971. Critics of the SST who feared a stacked deck were given some cause for concern when, on January 14—before the second CTAB meeting, and certainly before the panel had reached any final conclusions—an article appeared in the Washington *Post* concerning a briefing given to DOT's SST director, William Magruder, by Frederick Henriques, a physicist who headed the CTAB panel. Henriques said that the exhaust particles from high-flying SSTs would remain aloft for more than a year, while those produced by subsonic jets would drift to ground level in a matter of minutes. Henriques reportedly expressed a preference for the SST flights—a statement that seemed to imply that it was environmentally preferable to have the exhaust products high in the atmosphere rather than down at the ground (in fact, the opposite is true). Henriques was also reported to have said that no climatological changes would result from endless numbers of supersonic flights in the thin, stable stratospheric air. The *Post* story concluded that, while Henriques had a solid scientific reputation, it was doubtful the Administration would have chosen him had he been convinced on the basis of earlier studies that SSTs were undesirable.

The CTAB panel was not initially very interested in or concerned about the role of the SST in stratospheric chemistry. The credit for forcing this issue into the open goes to panel member Joe Hirschfelder, a brilliant and influential theoretical chemist from the University of Wisconsin who had been involved in analyzing the effects of the Bikini bomb test. Early in the panel's deliberations, Hirschfelder began to feel they were getting a bit of a snow job. He felt that too much of the information was coming from Boeing and that there was too little contact with government and university researchers who were experts in atmospheric chemistry. He was asking all the right questions (he even raised the question of nitrogen chemistry), but he wasn't getting any good answers. As it became clear that he wasn't going to get any from this group, Hirschfelder started lobbying for a meeting that would bring together researchers familiar with stratospheric chemistry. His campaign was not welcomed with noticeable enthusiasm by the rest of the panel. One observer who attended a panel session where the matter was discussed recalls that they seemed inclined to dismiss the chemistry problem out of hand. But Hirschfelder was angry. "I was going to blow my top," he recalled. Panel chairman Henriques was faced with the choice of having Hirschfelder blow his top or letting him organize the meeting. The latter choice seemed clearly preferable, so it was decided that the panel would gather for a scientific session in Boulder, Colorado, on March 18 and 19.

For Hirschfelder, the operative word was "science." He felt that the panel hadn't been coming to grips with the scientific issues, and he wanted an intensive workshop meeting where experts in the field of stratospheric chemistry could really thrash out those issues. He started making phone calls to various chemists at universities across the continent. Those who were on the receiving end of his calls found him extremely agitated about the situation, and most of them agreed readily to attend his meeting.

But when he called Harold Johnston at the University of California in Berkeley, he got a flat no. Johnston, who was, as a result of this call, to become a central figure in the SST debate, is a leading American physical chemist, a writer of books that are classics in the field. He was on sabbatical leave in residence at Berkeley that year, trying to catch up on his research and reading. It was going to be a nice, quiet year.

"I wasn't interested in SSTs," Johnston recalled later, rather drily. "And that year, I was declining all invitations to go anywhere or make any speeches." Johnston, who had started his scientific career during the war as a meteorologist in a chemical warfare research group, not only wasn't into SSTs, he wasn't even into the stratosphere. At that time he was unaware of Paul Crutzen's work on NO_x in the stratosphere. Much of Johnston's work had been on a very down-to-earth problem—the chemistry of smog. (Nitrogen oxides from car emissions help to *produce* ozone at ground level. A constituent of smog, ozone destroys rubber, including rubber tires—a subtle retribution. Less subtly, it also poisons human beings.) But Johnston had just finished writing a book on ozone and had spent twenty-five years studying the kinetics (speed of chemical reactions) of nitrogen oxides, and Hirschfelder unabashedly threw all of this up to him. "Joe was twisting my arm," Johnston said. "He wouldn't take no for an answer, so I said, all right, I'll go." And then Johnston went to check some books out of the library to refresh his memory on the stratosphere.

The scientists gathered in Boulder on the morning of March 18, about a dozen of them invited from the universities, along with scientists from the National Oceanic and Atmospheric Administration (NOAA), the National Center for Atmospheric Research (NCAR) and, of course, the members of the Commerce SST panel. Half a continent away, members of the House of Representatives were also meeting, and the SST was on their agenda too.

The Boulder meeting was a tense affair, marked by several rather bitter personality clashes. One that many of the participants remember vividly involved Jim McDonald who, on the afternoon of Thursday, March 18, outlined his concerns about skin cancer. Earlier in the day, Will Kellogg had noted in his opening remarks that, while he had questions about the medical aspects, he found McDonald's argument "generally convincing. Certainly it *deserves serious consideration,* though I think many of us tended to poo-poo it when we first heard about it." (Emphasis Kellogg's.)

McDonald nevertheless had a rough time when he gave his paper. He was berated by Harriet Hardy, a member of the panel and a professor of medicine at Dartmouth Medical School, who attempted to level him with the non sequitur that lung cancer was a much more serious problem than skin cancer.[1]

During his presentation, McDonald was subjected to a barrage of heckling and interruption by Arnold Goldburg, chief scientist of Boeing's SST Division. One observer remembered it as "real criminal-lawyer stuff." Goldburg kept demanding to know how McDonald *knew* ultraviolet radiation caused skin cancer. So scientists had done some tests on rats. Who cared about rats anyway? There were already too many rats in the world. Had scientists ever done these tests on humans? At least one participant at the meeting privately observed that he could think of a prime candidate for such tests.

"I have never seen any person at any scientific meeting so abused as [McDonald] was during the course of presenting his paper," Johnston said.

Goldburg does not characterize the exchange in quite the same way, though he acknowledges a gift for sarcasm. But feelings were certainly running high at the meeting. "It was a very emotional time," Goldburg said. What he recalls is challenging McDonald strongly about the theory on which the predictions of ozone depletion were based.

That theory involved water vapor emissions from SST engines.

---

[1] Hirschfelder says that Hardy was not really as unsympathetic to the skin-cancer problem as she seemed. It was just that she felt there were not enough good statistics on skin cancer. "Harriet was sympathetic but she didn't have any figures and everybody was trying to pin her down. And she was saying there's lots of good ways of treating skin cancer." Hirschfelder himself now has to use an ointment to treat pre-malignant keritosis, a condition which, if not treated, can lead to skin cancer. "So I'm more sympathetic now," he says.

At the time, most scientists, including McDonald, were worried about ozone depletion caused by HO_x from the water vapor, which later proved not to be the major problem. (No one was talking about NO_x in this context. McDonald had thought about NO_x, but dismissed it because he had not recognized the importance of the chain reaction in destroying ozone.)

Ironically enough, the first calculation of ozone reduction due to SSTs—using the same water-vapor theory that McDonald used—had actually been done by a Boeing scientist in 1970. Preliminary estimates by Halstead Harrison, then of Boeing's Scientific Research Laboratories, suggested large ozone reductions—reportedly around 20 per cent—but this was later characterized as a rough "back of the envelope" calculation. Boeing did not make these early calculations public, and Harrison's work would be referred to as the "secret Boeing report" in a congressional debate on the SST. Harrison did more detailed calculations and revised the ozone-depletion figures down to 3.8 per cent; this estimate was published in *Science* magazine in November 1970.

At the Boulder meeting, Goldburg became obsessed with recent measurements that indicated that ozone had been increasing in the atmosphere and that water-vapor levels had also been going up at the same time. This was not what the water-vapor theory predicted would happen, and Goldburg clearly believed these data to constitute a virtual death blow to McDonald's ozone-depletion calculations. But the experts in stratospheric chemistry did not feel Goldburg's conclusion to be warranted by these very limited data.

Shortly after the Boulder meeting, John Swihart, Boeing's chief engineer for the SST, wrote a letter to *Aviation Week and Space Technology* also arguing that McDonald's calculations were contradicted by measured ozone and water-vapor data, that the opponents of the SST knew this and "still perpetrated the 'Big Lie' on the American public. . . . To have publicly condemned the SST by use of a simple model which produced answers in direct conflict with the measured data is, at the least, scientifically dishonest if not treasonous." (Swihart had not attended the Boulder meeting, and his remarks were apparently based on the report of the meeting prepared by Goldburg—a report that, incidentally, ignored Harold

Johnston's new theory that NO_x was much more important than HO_x in destroying ozone.)

The "Big Lie" theme was a popular one among SST supporters when referring to those who opposed the development of the aircraft on environmental grounds. In addition to Swihart's letter, *Aviation Week* ran an editorial by Robert Hotz that charged that the use of the skin-cancer theory was a "cruel and cynical exploitation of the ecological hysteria" and concluded that this "Big Lie" was "aimed at creating a wave of false sentiment at a crucial voting time."

Julius London, a meteorologist at the University of Colorado who had published the data on the increases in the ozone layer, was provoked by the charge of scientific dishonesty to write a letter to *Aviation Week* in reply. He said that both Hotz and Swihart "fell trap to their emotions and failed utterly to do their homework before sounding off." Swihart, he said, either had never seen the ozone data to which he referred or, worse, hadn't the faintest idea what it meant. He then pointed out that several independent estimates of the stratospheric effect of SST exhaust products (including one by Harrison "who at the time . . . was and, I hope still is, employed by Boeing") indicated a reduction in ozone that was "certainly cause for concern." According to London, the letter was not published.

Boeing officials could not seem to overcome the antipathy bordering on contempt that they felt for McDonald. One referred to him as "that UFO expert." Goldburg, who had heard McDonald talk on UFOs, could not understand why the National Academy would associate itself with a man who engaged in this "highly questionable kind of scientific activity."

Though McDonald had, in his National Academy study, dismissed the charge that SST water-vapor emissions would alter the climate, though he had emphatically reiterated these conclusions both in his March congressional testimony and in his presentation at Boulder, though he had demonstrated he was not personally or politically motivated to bring down the SST at the expense of scientific truth, "that UFO expert" was rarely given any credit for this by Boeing. In an article in *Astronautics and Aeronautics,*

Goldburg discussed the 1966 National Academy study of water-vapor effects on climate—noting that its conclusions "are standing the test of time very well"—without acknowledging McDonald's authorship of this study in the text of the article (though he does name McDonald in the references). On the other hand, later in the article, however, Goldburg prominently mentions McDonald as the originator of the skin cancer/SST link.

McDonald died tragically by suicide in mid-1971. There was the inevitable speculation that the death was tied to the ozone controversy, but there is no proof of this. Today, McDonald has many defenders among those who were his colleagues. They rise protectively to his defense, for they do not want him remembered as a UFO nut. J. G. Charney, a meteorologist from the Massachusetts Institute of Technology (where McDonald had also taken a meteorology degree), defended McDonald's unusual views on UFOs as evidence of his fearless pursuit of the truth. London, who considered McDonald a good friend, agreed. "He was a broad individual. He did get on kicks. But he was a concerned individual and this is what people didn't realize—the depth of his concern."

London, who was one of the speakers at the Boulder meeting, was having troubles of his own there as a result of an incautious statement he had made in a report published the year before. The report, ambitiously entitled "Man's Impact on the Global Environment: Report of the Study of Critical Environmental Problems" (called SCEP, since no one can ever remember its full name), was the result of a wide-ranging interdisciplinary workshop held in July 1970. Julius London, one of about seventy participants, was a member of the working group on climatic effects. The environmental impact of SSTs was only one of a large number of issues being studied, and it fell to the climatic-effects group, which was dominated by meteorologists and lacked a strong expertise in the chemistry of the upper atmosphere. Meteorologists were not really the best people to be considering pollution of the upper atmosphere by SSTs, at least insofar as it involved questions of chemistry. Strange as it may seem to the layman, meteorologists and stratospheric chemists had long gone their separate ways. There was little dialogue and even less love lost between the two groups; their relationship, such as it was, was characterized primarily by a kind of jurisdictional warfare over the earth's atmosphere.

Meteorologists generally regarded transport—that is, the movement of air—as overridingly important in understanding the atmosphere, and they often rather contemptuously dismissed the importance of chemical reactions, which tend to be more localized phenomena. Chemists, on the other hand, tended to regard the effect of air motions as more of a nuisance than anything else and preferred to disregard them as much as possible. This provided each camp with a handy scientific scapegoat when they ran into things they couldn't explain about the atmosphere. The chemist would explain what he could in terms of chemical reactions, and what he couldn't explain he would blithely dismiss as being due to air motions. The meteorologist would do the reverse.

Nevertheless, the SCEP meteorologists did consider the question of SST impacts on the ozone layer. This was largely because of the involvement of Julius London who, interestingly enough, had long had an interest in bridging the gap between meteorologists and chemists. London was an expert on measurements of the ozone layer and had himself earlier done calculations of ozone depletion due to water vapor. On the basis of his work, the SCEP report concluded that ozone reductions due to SST operations should be "insignificant."

But it was not the water-vapor work that got London into trouble—instead it was the statement that the role of nitrogen oxides (NO_x) in stratospheric chemistry "may be neglected."[2] This conclusion takes up only a single sentence in a report of more than three hundred pages; yet it was fated to become much-quoted and much-criticized. It was one of those sweeping, categorical statements about nature that—since nature pays no attention at all to these learned treatises—often get scientists into trouble. As London ruefully acknowledged: "You know, once in awhile you put things in writing that you wish just were not there."

London had dismissed the NO_x problem in the stratosphere on the basis of calculations that showed NO_x to be unimportant in the region above the stratosphere (called the mesosphere). He was not

[2] Ironically, it turns out London may well have been right, but for the wrong reasons, as we shall see later.

really justified in making this extrapolation, but nevertheless, when the question of NO_x emissions from SSTs was raised at the SCEP meeting, he rejected it. The question was raised by Lester Machta of the National Oceanic and Atmospheric Administration. Both he and London rather sheepishly accept the blame for putting the NO_x problem temporarily off the tracks. Machta had come to the SCEP meeting with some hard figures for the amount of NO_x the SSTs would put into the stratosphere. When he raised the question, he was trying to point out to London that this amount of NO_x would add considerably to the natural NO_x in the stratosphere. But for some reason, Machta and London just weren't communicating, and London did not latch onto this concept. He continued to reject the NO_x problem in light of his earlier calculations based on natural NO_x.

London was also aware of Paul Crutzen's work on the role of the NO_x chain in destroying stratospheric ozone, but London had a "gut feeling" that this wasn't important and so ignored it. (Crutzen, who had been trained as a meteorologist, was still feeling rather insecure about his chemistry at the time, and he was not pushing his ideas in scientific circles as vigorously as he might otherwise have.)

Johnston was extremely annoyed by London's dismissal of the NO_x problem and (when he later found out about Crutzen's work) deeply suspicious about SCEP's lack of reference to Crutzen, seeming to regard this as verging on suppression of scientific information. He and London clashed head-on at the Boulder meeting, where Johnston argued that NO_x would reduce stratospheric ozone a great deal more than London had calculated.

Johnston is a slight, white-haired man with a professorial manner, a seemingly introverted type who at first glance strikes many people as a most unlikely candidate for the role of scientist activist. There seems, at times, a fragility to him. He has incongruent qualities—for example, a soft, quiet voice that does not seem entirely consistent with his rapid-fire, intense, even passionate manner when he speaks on a subject that exercises him. And what London had said definitely exercised him.

To Johnston, the problem stemmed in large part from the long-standing division between chemists and meteorologists, and perhaps not surprisingly he thought that most of the fault lay in the meteorologists's camp.[3]

The exchange between Johnston and London had a considerable impact on both the participants. Johnston's recollection of the incident conveys the impression that he felt himself surrounded and threatened. He recalls being "jumped on in the meeting and in the corridors. They just let loose emotions." (Others, however, do not remember any such ganging up.)

London has not forgotten the incident either. "It was a charged kind of atmosphere," he says. The experience was clearly a painful one for him, and he seems to feel that the less said about it the better. When pressed, he relates the tale blandly, saying only that Johnston "kind of pointed the finger" at SCEP.

Goldburg, the man from Boeing, does remember the exchange. "Johnston really sliced London up pretty good. He has very good forensic skills." Goldburg knew London and believed him to be a dedicated SST opponent, but it was his first exposure to Hal Johnston.

London had never heard of Johnston either, and was rather taken aback to find himself on the receiving end of what seemed a very emotional outburst. Today, he speaks very highly of Johnston, but at the time the incident definitely rankled. London began to take Johnston seriously only after a conversation in which he bitterly discussed the incident with Schiff, who assured him that Johnston was neither a crank nor a kook but, in fact, a top expert in the field of nitrogen oxides. Nevertheless, the incident was to begin a period of seriously strained relations between Johnston and London that would not be patched up until nearly two years later.

The first day of the Boulder meeting ended on a rather dramatic note. Late Thursday afternoon, word began to spread throughout

[3] The meteorologists got in a few tart comments of their own. A year after SCEP, a follow-up meeting was held, called the Study of Man's Impact on Climate, or SMIC. The SMIC authors devote a good deal of attention to the nitrogen oxide/ozone/skin-cancer problem. They refer to Johnston's calculations, but note that his results are based primarily on chemistry and add: ". . . the problem of stratospheric ozone distribution is very complex and must be studied in an atmospheric model in which proper treatment is given to dynamical processes [air motions] before any definitive results can be considered reliable."

the room that Congress had just killed the SST project. In actual fact, the House of Representatives had voted 215 to 204 to deny Boeing $290 million in federal funds for the following year to continue work on the two prototypes that were then being built. If federal money was denied the project, it would almost certainly die, since private capital was nowhere to be found.

The announcement had an electrifying effect on the scientists at Boulder. Most were not particularly well versed in the ways of Congress, and they were largely unaware that the House vote did not necessarily end the political fight over the SST—that the Senate had yet to vote on the issue. An aura of anticlimax settled over the meeting, but the group decided to press on, since the ozone problem was global in scope and the Soviet and French-British SST programs were alive and well.[4]

Goldburg took the news hard. A good part of his life had been wrapped up in the SST project, and its demise was demoralizing to him, particularly since he thought it a "bum rap." There was to be a party that night at Julius London's home, and Goldburg was undecided about going; he felt that many of the people who would be there had been responsible for the SST's death, and he did not particularly relish the prospect of engaging them in small talk over cocktails. But staying away seemed a "purposeless gesture," so he went.

Johnston was in an emotional turmoil over the events of the first day, "distressed by the ferocity of some people's comments to me and by the tenor of Goldburg's interruptions and statements." Johnston stayed up all night to do more detailed calculations on the NO_x problem and prepare a presentation for the next day's sessions. When the delegates reassembled on Friday morning he had photocopies of his handwritten paper ready for them. In it, Johnston calculated that two years of operation by five hundred SSTs would cause global ozone reductions between the "serious" level of 10 per cent and the "catastrophic" level of 90 per cent, allowing ultraviolet radiation at wavelengths "never before encountered during the evolution of man" to reach the earth's surface.

[4] Goldburg was concerned whether the SCEP authors would request that the Soviet Union stop flying SSTs. He said he was told this could not be done because there was no proof an SST fleet would have any effect on the environment.

Arthur Westenberg of Johns Hopkins University was also concerned about NO_x—he had, in fact, done some computer calculations of ozone depletion prior to the Boulder meeting—and he too
presented his thoughts on the matter at the meeting on Friday
morning.

By midmorning many of the scientists were getting restive. The
formal presentations were still going on, and there seemed little
prospect for the gloves-off scientific bull session that Joe Hirschfelder had promised. Some of the scientists began to feel slightly
railroaded. It was at about this time that a cryptic note from
Hirschfelder passed among the invited university scientists. "Meet
you in the men's washroom in ten minutes," it read. Intrigued, the
small group of about a dozen scientists trooped off during the
next break to find out what Hirschfelder had on his mind.

Hirschfelder was determined to have his scientific workshop, and
during the washroom summit meeting he announced to those as-
sembled that they were going to find a quiet room with a black-
board and sit down to do what they had come to do. They packed
into a small seminar room, distributed themselves on chairs and
desks, and proceeded to discuss the problem. Initially the discussion
revolved around the ozone/water-vapor problem. Little was said
about nitrogen oxides.

Johnston sat quietly for a time. Then he lashed out at his colleagues for their inattention to what he'd been saying about NO_x.
All this talk about water vapor completely missed the mark, he told
them angrily. Any freshman chemist could see that the real problem
was NO_x.

The chemists in the group were stung into immediate rebuttal.
There was an element of wounded pride in this; they were not,
after all, freshmen chemists, and they did not take kindly to being
compared to them—and unfavorably, at that. Several of the scientists in the group had been deeply involved in making laboratory
measurements of the relevant chemistry. Fred Kaufman of the University of Pittsburgh was an expert on the water-vapor/ozone reactions being discussed, and Schiff had been the first to measure the
speed of the reaction between nitric oxide and ozone.

The atmospheric physicists, particularly Mike McElroy of Harvard, were also critical. Johnston had virtually ignored atmospheric

motions, which the physicists believed would dominate the situation and make the conclusions based on chemistry largely invalid.

Although there may have been some interdisciplinary rivalry here, and perhaps some sheepishness, the attempt to shoot holes in Johnston's theory was, at the same time, standard operating procedure with a new scientific hypothesis. It is the way science is done.

The group thrashed around with the problem for some time, raising objections, speculations, and a variety of yes-buts. The most important unanswered question was how much NO_x was already in the stratosphere naturally. If the SSTs were going to add only a tiny fraction of the natural amount, they could not be a serious problem. But no one had measured NO_x in the stratosphere, and no one knew how plentiful it was naturally.

Johnston became impatient with his colleagues. Convinced he had caught a very serious problem only barely in time and anxious to get his concern out in the open, he tended to dismiss many of the caveats being thrown at him. But the simple fact was that the scientists had very little hard data to go on. What finally emerged from the session was the view that Johnston had indeed identified a potentially significant threat but that the scientific unknowns were too great to reach any conclusions. They reported this back to the panel with a recommendation that a research program be started to answer the questions. It was a middle-of-the-road approach—and Johnston clearly regarded it as rather gutless—but the unanswered questions made the group nervous.

When the entire group reconvened on Friday afternoon, Johnston moved that the group repudiate the SCEP statement that the effect of NO_x in the stratosphere "may be neglected," and he recommended a thorough study of the problem. "The motion was brushed aside," Johnston says. "I do not recall any special support from the other participants, and someone asked me if I believed in flying saucers."

For his part, Goldburg got up to argue that there were not enough data to determine whether or not SSTs would harm the environment, that a more detailed data base would be needed before such a determination could be made. He vigorously approved the idea of putting more research funds into the effort, something the cynics in the audience chose to interpret as a subtle form of bribery, an interpretation Goldburg both resents and denies.

In the end the panel did recommend a program of research and measurement to resolve the unanswered questions. In his letter to Peter Peterson, then Secretary of Commerce, panel chairman Fred Henriques said: "At present, there is insufficient technical data to render a decision on whether or not a large number of flights of supersonic transports would cause significant environmental hazard to mankind."

On March 22, about a week after the Boulder meeting and two days before a crucial senate vote on the SST, the Department of Transportation's SST Advisory Committee met in Washington. This committee was composed largely of government-employed atmospheric scientists, and its function was to help DOT set up an SST environmental research program. Its chairman was Fred Singer, a physicist who was then also deputy assistant secretary of the Department of the Interior. At the time that he accepted the chairmanship of the SST committee, Singer had stated that he was not a supporter of the SST program and had not previously taken a stand for or against the project. He accepted the position, he said, "because I want to make sure that scientific data are not misused, either to support the SST program or to oppose it."

Singer did not reject the environmental concerns out of hand. In congressional testimony given shortly after Jim McDonald's, he used carbon monoxide from automobile exhausts as an example of a man-made pollution problem (unrelated to SSTs) that nature, for reasons unknown, appeared to be coping with. "We seem to be safe, not because we were smart but because we were lucky. We can't always be lucky. That is why we must take all possible environmental effects, including those of the SST, seriously. . . ." Nevertheless, Singer was extremely dubious that the environmental impact of SSTs would be significant. While be believed McDonald's skin-cancer calculations to be exaggerated, Singer still urged DOT to consider the question seriously. And he concluded his congressional testimony (in early March, before the Boulder meeting) by saying: "There is no question that the SST is going to release some pollutants into the atmosphere, but it is doubtful whether they will be of any significance." He said that the decision whether to proceed with the SST project should be based on economics and national priorities "with the environmental effects having a very small weight indeed. If the SST is going to be turned down, let's be sure that it is turned down for the right reasons."

At the March 22 meeting, Singer managed to anger Lester Machta sufficiently that Machta stormed out of the room. Both men remember his departure but, other than that, they recall the incident very differently.

According to Machta, who wrote notes on the meeting shortly after it occurred, Singer presented the advisory committee members with a statement analyzing—and casting doubt on—McDonald's thesis. Singer was seeking endorsement and Machta says he asked if the statement was to be made public. Told that it was, he got up from his seat and left the room. "I walked out because I felt there was absolutely no possibility of getting a fair deal on the environmental side," Machta recalled in an interview. "He was twisting our arms so vigorously that there wasn't going to be a fair hearing."

Singer does not share this view. He says his attitude was that McDonald might have something, but that there were things he (McDonald) hadn't taken into account and that the matter had to be properly studied. (However, Singer did believe that "the economic factors against the SST were so strong that, in relation to them, any environmental influences would have to be negligible.")

Singer remembers Machta walking out of the meeting. "We were all surprised. I think it was in response to the fact that the committee felt—most of the committee agreed—that there were things in the McDonald story that didn't hang together and would have to be studied." Singer says he does not remember any public statement coming out of the committee, "nor do I think we considered one."

In fact, no public statement was issued, but a letter, which Singer signed as chairman of the committee, did appear in the Washington *Post* shortly before the Senate vote. Machta, who saw it as yet another attempt to counter the environmental anti-SST position, says he took "very great exception" to it. Singer, however, said it was a personal statement—"because I certainly did not submit anything to the committee for clearance"—and added that he had signed himself as committee chairman only for purposes of identification.

Hal Johnston was not content to let the SST matter rest in the hands of the CTAB panel. Shortly after he got back to Berkeley, he made a telephone call that would entangle him further in this issue in which, just a few weeks before, he had had no involvement and little interest. He called the White House, asking to speak to Edward David, then the science adviser to President Nixon. Johnston wanted to alert the White House to what had happened in Boulder.

The swiftness with which Johnston moved into the political arena provides a striking contrast with Sherry Rowland's action in 1974 of keeping a low profile on the fluorocarbon problem for nearly six months. It would have been less surprising if the roles had been reversed. By 1974, political activism by scientists had gained, if not wholehearted approval, then a certain grudging acceptance within the scientific community, but in 1971 such activity was still considered vaguely disreputable. Given the times, Johnston's was an act almost guaranteed to incur the disapproval of his scientific colleagues and spark cynical speculation as to his motives. This fact did not escape Johnston, who, even today, relates the circumstances of the call defiantly, offering no apologies. It is clear that, rightly or wrongly, Johnston felt he had a responsibility to tell the people in charge about a situation that deeply alarmed him. He was convinced it was possible "they were leading the world right into an absolute disaster." Fumbling for the words to explain his motivation, he finally shrugged his shoulders and said: "You go through your life kind of doing your pure science and teaching your students, and then, all of a sudden, if it turns out that what you're doing has an impact, you've got to talk about it. You've got to say it. . . ."

Johnston did not get a welcome at the White House. He never did get through directly to David, and it was about a week after his initial call before a member of David's staff got back to him. By then, the Senate vote—which would reject the SST by 51 to 46—was in progress. The staffer pointed this out to Johnston, who said he was concerned about the long-term implications of the problem and not just about the Senate vote on two Boeing prototypes.

The staffer told Johnston to put it in writing and they'd think about it.

"Following that, I made my one tactical error," Johnston said. "I stated the thing in such strong language. You know, scientists are always supposed to be very calm, dispassionate, neutral. But in my letter to him, I laid it on very heavy as to what kind of biological effect you might have if you were reducing ozone by a factor of two."

For example, he said this would lead to a serious, widespread problem of ultraviolet-induced blindness (a phenomenon that would be similar to snow blindness, caused by the reflection of ultraviolet light by ice and snow). Johnston wrote that "all animals of the world (except, of course, those that wore protective goggles) would be blinded if they ventured out during the daytime." This statement later received rather sensational treatment in the press and was even featured in a children's comic strip called *Frontiers of Science* published in April 1972. Johnston stands by the statement, but he acknowledges that it probably did more harm than good. "Many people, when they tried to show how irresponsible I am, quoted this blinding thing. I know that when I wrote that, I was really trying to light a fire under the stubborn mules who weren't listening. They were taking a ho-hum attitude—they didn't believe a word of it, and I was trying to catch the attention of those in charge."

Johnston could not have known just what he was getting himself into; but in fact he had blundered right into the middle of a dandy political war that had been raging for more than two years over whether to cut bait or fish on the increasingly expensive Boeing SST project. President Nixon was pro-SST and had taken a personal interest in the battle. From all accounts, dissidence within the executive branch was neither encouraged nor rewarded. The day before the crucial congressional vote on March 18, the White House had released an indignant statement protesting SST opponent Senator William Proxmire's use of the skin-cancer issue, saying it was a "shocking attempt to create fear about something that is simply not the fact."

Now, here was Johnston telling them it *was* a fact. No, the pro-SST White House was not anxious to hear his story, and rumors soon started circulating in the scientific community that the matter had reached the "highest levels of the White House" (a phrase that was to gain some notoriety during the Watergate era), where there reportedly was talk of "getting this guy Johnston." Johnston himself claims, not without a certain relish, to have heard from one source that news of the heated debate in Boulder had rapidly reached the ears of John Ehrlichman—then the top domestic adviser to Nixon—who was reported to have said that if these scientists couldn't agree with one another, the White House wouldn't pay any attention to any of them. This was later confirmed publicly by Hubert Heffner, former deputy director of the White House's Office of Science and Technology. In a speech, Heffner suggested that when scientists reach opposite conclusions based on the same evidence, the public and government officials tend to disregard their views on political issues. He is quoted as saying that such lack of public confidence enabled Ehrlichman to dismiss Johnston's calculations by saying, "No one believes scientists anyway."

The ozone/skin-cancer "scare" would in hindsight receive much of the blame (or credit, depending on your point of view) for killing the U.S. SST program; indeed, an impression lingers that it was almost single-handedly responsible for bringing down the billion-dollar technological Goliath. But this is not so. The ozone issue was more in the nature of the straw that broke the camel's back. The environmental charges against the SST involved a polyglot collection of concerns about the sonic boom, aircraft noise at ground level, pollution of the lower atmosphere and possible climatic changes, as well as the threat to the ozone layer. Most of these other concerns emerged very early in the fight and, at least initially, received more attention than ozone. As a public issue, the ozone/skin-cancer problem was in fact an eleventh-hour arrival on the scene in 1971, and it was useful as a political tool only in the last three months—though admittedly the most important months —of a fight that had been going on for years. It and the other environmental questions were often eclipsed by the fierce assault that was launched against the SST on economic grounds—a subject that one suspects was at that time much nearer and dearer to the hearts of congressmen than the environment.

It is true that during the last half of 1970, Senator Proxmire, who had been fighting the SST since 1963 largely on economic grounds, did embrace the environmental cause. The New York *Times* reported: "In a comment that reflected the changing SST politics, Senator Proxmire said today, after summarizing his economic arguments against further development, that the paramount reasons for postponing the SST involved the environment." (A gas station attendant in Seattle, Boeing's home base, would later question whether Proxmire was "Peking's man.") Proxmire seized on the skin-cancer issue quickly. Shortly after Jim McDonald gave his testimony in Congress at the beginning of March, Proxmire wrote to Gio Gori, then associate scientific director of the National Cancer Institute, asking for an evaluation of McDonald's conclusions. Gori responded quickly with a brief report that basically supported McDonald's position and gave even larger estimates of the additional number of skin-cancer cases to be expected if eight hundred SSTs went into operation. (One scientist who was privy to some internal discussions said Gori "deserves quite a bit of credit for being forthright on this score.")

Proxmire released Gori's report at a press conference on March 17, just one day before the House vote that marked the beginning of the end for the U.S. SST. Proxmire claimed that the Administration had tried to muzzle Gori, but Gori publicly denied this.

The skin-cancer issue certainly made headlines in mid-1971, but it seems clear despite this that economics played the dominant role in defeating the SST in the end. The events of one week in mid-May seem to bear this out. On May 12, the House of Representatives reversed its earlier vote against the SST in a stunning upset orchestrated by then minority leader Gerald Ford. Ford actually managed to have the bill authorizing cancellation funds amended to read "program renewal." The Senate vote on the bill was scheduled for a week later.

During that week, several things happened. A copy of Johnston's paper was leaked to a small California newspaper, the Newhall *News,* apparently by someone who had been asked by the White House to evaluate Johnston's work. Johnston, who denies leaking the report to the press himself, said that the large California papers then started leaning on Berkeley, and he was reluctantly persuaded by the university to sanction a press release. The news obtained what an irate Boeing official characterized as "one-inch screaming headlines" in the western press, and on May 17, two days before the Senate vote, the more sedate New York *Times* also published a story about Johnston's ozone calculations. It quoted Johnston saying that five hundred SSTs flying seven hours a day would take less than a year to deplete half the ozone layer. The story also noted that "Senate opponents of the SST indicated that they were aware of Dr. Johnston's study but that they had not intended to introduce the subject so soon before a vote that they already expected to win."[5]

But another important event also occurred during that crucial week in May. William Allen, chairman of the Board of The Boeing Company, announced that it would cost at least $500 million and maybe as much as $1 billion to revive the SST prototype project, which had been winding down since March—and that was just to continue work on the two prototypes; further government money would be needed to start production of the aircraft, he said. SST supporters swallowed hard when they heard this. There were large expenses involved in terminating the project, and they had been arguing that it would cost *more* because of these termination payments to stop the SST project than to keep it going. Allen's figures cut the ground out from under them; termination payments certainly weren't going to run into the billion-dollar ball park. The fight went out of them then. As the Senate vote approached, the White House conceded defeat through presidential press secretary Ronald Ziegler and placed the blame squarely on Boeing's refusal to modify Allen's figures. Senator Warren Magnuson, an SST supporter, did not offer much of a defense; in his brief remarks from the Senate floor that afternoon, he bleakly "noted the burden of costs."

The protechnology lobby likes to blame those they choose to call "eco-freaks" for the demise of the Boeing project, but the fact is that the ozone problem was only one of many contributing fac-

---

[5] Late in April, Johnston had discussed the problem with former presidential science adviser George Kistiakowsky and, according to Johnston, Kistiakowsky verbally discussed the matter with Senator Proxmire and Senator Henry Jackson when the SST issue revived in May.

tors and probably not the dominant one. It might not be true today, but in 1971, concern about the ozone layer did not possess nearly as much clout as the horrifying billion-dollar price tag that was dangled before congressional eyes.

While the political wars were unfolding, the scientific community was also coming to grips with the ozone issue. At the beginning of April, shortly after Johnston returned to Berkeley from the Boulder meeting, he prepared a scientific paper in which he calculated quite large ozone reductions—average global reductions ranging from 3 to 23 per cent, with 50 per cent local reductions near zones of high SST traffic.

Johnston submitted the paper to *Science* magazine, which then sent it out to various scientists to be reviewed. The National Academy of Sciences convened an ad hoc committee on ozone and also sent the paper out for comments and evaluation. (This committee would conclude, a considerable time later, that Johnston "has done a service in bringing this problem into full view and stirring the needed debate within the scientific community.")

The White House also took a considerable interest in the paper, forwarding an early version to scientists at the National Bureau of Standards with a request to determine whether it had any merit. The White House pointed out to the NBS scientists that the final Senate vote on the SST was approaching and that it would be a shame for the paper to influence the vote if it was in fact not correct. Those on the receiving end of the call interpreted it, perhaps wrongly, to mean that the White House was asking them to see if any flaws could be found in the paper sufficient to delay its publication until after the Senate vote. After a flurry of telephone calls over one weekend, the NBS scientists reported back to the White House that the paper seemed basically all right.

The scientists who received Johnston's paper from *Science* and the National Academy agreed that he had identified a potentially serious problem and accepted the fundamentals of what he was saying. But they could not unquestioningly accept the magnitude of the effect Johnston was predicting; they felt that his exact figures could not be supported because the relevant data were lacking. In particular, they criticized his assumption that there was little NO_x

in the stratosphere already. No one knew whether this was so or not.

But what alarmed the reviewers most was the tone of the paper. It was not the studiously dispassionate document such papers are supposed to be. In their view, it contained too many emotional and political statements. They were rather horrified by the statement about animals having to wear goggles, and by the concluding sentence, which read: "In view of the eye-destroying, deadly nature of the radiation shielded from the surface of the earth by ozone, the prospect of the destruction or the major reduction of the ozone layer should be regarded as a matter of utmost worldwide concern." This was distinctly to be frowned upon. References to practical or political ramifications are now more acceptable in scientific papers, though still carefully policed, but back in 1971, the peer group thought Johnston had gone too far. The reviewers were also alarmed by the rumors of White House hostility toward Johnston and were concerned that if scientists made statements that were too extreme, soon no one would take the SST issue—or anything else scientists said—seriously.[6] So the peer group turned the paper back for a rewrite. The second version concluded by saying that the speeds with which nitrogen oxides would react with ozone were well known "and they firmly imply catastrophic reductions of stratospheric ozone if (the nitrogen oxides) reach concentrations deduced by SCEP as the expected worldwide stratospheric average." This version also went back to Johnston. The final version, which appeared in *Science* in August, was considerably toned down. Though Johnston was still predicting ozone depletions as high as 50 per cent, he noted that the purpose of the paper was not to give precise ozone-reduction figures, but to point out that the NO_x problem was much more important than the water-vapor problem that had been receiving most of the attention. Gone were the references to eye-destroying radiation; Johnston concluded the paper more meekly by saying:

"Just as the SCEP report incorrectly discounted NO_x and the SST planners for several years overlooked the catalytic potential of

---

[6] One of the rumors had it that President Nixon had been so incensed by new reports of Johnston's calculations that he ordered aides to investigate the possibility of having Johnston's research funding cut off.

NO_x it is quite possible (and in fact highly probable) that I have overlooked some factors and the effect of NO_x on the zone shield may turn out to be less or greater than that indicated here."

Five years later, a major research program costing more than $21 million would arrive at the same basic qualitative conclusions that Johnston did.

Johnston's opponents received much the same kind of treatment at the hands of the reviewers. In response to Johnston's paper, Arnold Goldburg wrote a letter to *Science* criticizing the Berkeley scientist for taking his extreme case of ozone depletion (50 per cent reduction) and "rush[ing] into headlines with it from New York to San Francisco three months before the paper is even in print. . . ." This was not, in Goldburg's view, "a constructive contribution to the goal of obtaining scientific decisions in the public interest."

Goldburg's paper was rejected by the reviewers, in part because they considered it too "personally vituperative."

In the months after the Boulder meeting, Johnston became engaged in a vigorous defense of his theory. He clearly felt himself in an embattled position, as an exchange of letters with Will Kellogg in Boulder demonstrated. Kellogg had previously written an article in which he had commented that "the general feeling of those I have talked to is that Johnston has overstated the effect [of SSTs on ozone] by a rather wide margin. . . ." Johnston took exception to this and wrote a long and rather technical letter to Kellogg in rebuttal. He referred to remarks Kellogg had made at the Boulder meeting with a comment that would later become familiar to anyone who engaged Johnston in debate: "I took detailed notes." (At most scientific meetings, Johnston can be seen meticulously recording in the minutest detail practically everything that transpires.) Johnston's disagreement with Julius London, who was also in Boulder, had continued to escalate, and Johnston was feeling much aggrieved by the "Boulder group." In his letter to Kellogg he said: "My impression is that very few people at Boulder have bothered to read with understanding what I have written, except in a hasty search to try to find weak spots. . . . Without doubt, there are some things in the 1971 reports that are incomplete and that will turn out to be wrong even though based on the best 1971 knowledge. However, I am getting very impatient with the Boulder group talking about my work as if I had deliberately or foolishly 'overstated the effect by a rather wide margin.' Please examine my actual articles. I think you will find that nowhere is there any exaggeration, overstatement or unfair selectivity."

However, Johnston ended the letter on a conciliatory note. He said he was not trying to embarrass Kellogg or London and added: "During the next couple of years it will be necessary for chemists and atmospheric scientists to work together to solve this problem. . . . At present, Berkeley and Boulder are at a stand-off; each feels wronged and insulted by the other. Is it possible for you and me to bring out our mutual complaints so that some of the misunderstanding can be cleared up? . . ."

Johnston says he does not, in retrospect, regret the controversy of 1971. At the time he thought the news stories about his early work were counterproductive, but now he's not so sure. He feels that unless issues like this are brought out into the open, the people in positions of power can ignore them. He firmly believes that without the publicity, the U. S. Government might never have set up the Climatic Impact Assessment Program (CIAP) to examine the stratospheric effects of SSTs. "The DOT said they had planned even back in 1970 to make a study but Congress turned them down in 1970 and probably would have turned them down again in 1971," Johnston said.

Among other things, Congress asked for an evaluation of the effect that SST operations would have on the amount and distribution of ozone and the impact that changes in the ozone balance would have on humans, plants, and animals. The Climatic Impact Assessment Program inherited its title from an earlier time when the main concern was climatic effects of the SST, but it was the ozone question that overwhelmingly dominated the CIAP study.[7]

The study would take three years, involve the efforts of some one thousand scientists from ten countries and cost $21 million in direct funds, plus about $40 million in indirect contributions from government agencies and universities. In the end, the seventy-two hundred

[7] A marked emphasis on the climate question by the head of the CIAP study, Alan Grobecker, would later cause considerable controversy. See Chapter Four.

pages of data would lead to the conclusion that Johnston had been right, in essence if not in every detail. A large fleet of hundreds of Boeing-type SSTs, using existing engine technology, would indeed be a threat to the earth's ozone layer. The CIAP estimates for reductions in the global-average ozone ranged from about 10 per cent to 20 per cent compared with Johnston's estimates of 3 per cent to 23 per cent.[8]

The CIAP study has been described as the most ambitious attempt at technological assessment ever undertaken. The scientists who took part in it viewed its accomplishments with considerable pride, which is why they were horrified and outraged when they saw the CIAP executive summary. The executive summary debacle started with a press conference in Washington on January 21, 1975.

[8] As we shall see later, continuing refinements in our understanding of stratospheric chemistry would, more than two years later, lead to a downward revision in the calculations of SST impact.

CHAPTER FOUR

The Great Executive-summary Snafu

SST Is Cleared on Ozone.

—Headline in Washington *Post*,
January 22, 1975.

The news reports out of Washington at first puzzled the CIAP scientists, then angered them. How it could have happened, they did not know, but there it was, the Associated Press story of January 22, 1975, saying these strange things.

"A three-year study dispels fear that the present fleet of supersonic transports will damage the earth's protective shield of ozone, the Department of Transportation said yesterday."

The statement was not exactly wrong, as far as it went. It was just that it entirely missed the point. The scientists were not really that worried about the effect of the existing miniscule fleet of first-generation SSTs. In January of 1975 this fleet consisted of a handful of French-British Concordes and Soviet TU-144s—a number that was, as one science writer acidly observed, "unlikely to represent a threat to anything except Britain's balance of payments."

What the CIAP scientists were concerned about—what they thought they had spent the last three years studying—was the *potential* effect on the earth's upper atmosphere of a large fleet of hundreds of SSTs, particularly the more advanced versions such as the one Boeing had planned to build. Indeed, the main thrust of the study was to determine what would likely happen if the French-British and Soviet programs expanded and if the U.S. plans for fleets of five hundred to eight hundred SSTs had proceeded unchecked. The result of this study was not reassuring, and this is what made the second paragraph of the AP story so alarming to the CIAP scientists.

"Dr. Alan Grobecker, who directed the study, said a U.S. fleet of high-flying planes would not have weakened the ozone shield either."[1]

The headlines were even worse. "World SST fleets said not to damage ozone blanket," intoned the New York *Times*. "SST is cleared on ozone," announced the Washington *Post*. "DOT study says the SST does not damage the ozone blanket," proclaimed the "CBS Evening News." Although some got the story right, these were the headlines that dominated coverage of the CIAP report across the nation.

Swiftly, and with a feeling of dismay, the scientists who had participated in the CIAP program began to realize that somehow the gist of the entire study had been turned around approximately 180 degrees. No matter that the study itself had essentially confirmed their fears about the effects of a large fleet of advanced SSTs; no matter that a thousand of their number from ten countries had spent three years and more than $21 million of the government's money to find this out. With a sinking feeling, they realized they would probably be retroactively branded alarmists and fellow travelers of the "disaster lobby"—all on the strength of misleading headlines.

Their fears were soon realized. A column in the San Francisco *Chronicle* by John D. Lofton, Jr., announced that "the facts are now in. The anti-SST people were wrong. . . . The Eco-Freaks and their allies got their victory. But the victory came at the expense of truth. . . . The supersonic transport was defeated by a sophisticated and well-orchestrated use of the Big Lie technique. . . ."

Another article, by Roscoe Drummond in the *Christian Science Monitor*, flatly stated that the SST threat to the environment and human health was "not only unproved, but proved to be wrong." He took Congress to task for having killed the SST on the basis of "ecological myths" and said that the "authoritative, conclusive, and

---

[1] Grobecker denies making this statement. As we shall see, it arises from a misunderstanding of something he did say.

reassuring" CIAP findings gave the green light for a resumption of the U.S. SST program.

Drummond's article was entered into the *Congressional Record* by Republican Congressman Robert McClory of Illinois, who said that the article reviews "the solid evidence which establishes the basic safety of the SST—and the groundless arguments which were employed by many of our colleagues in destroying this useful development in air transport." McClory declared that unfounded fears, wild speculation, and emotional appeals to prejudice had inflicted permanent damage to the nation's economy and aviation leadership.

Nor were the CIAP scientists the only ones taken to task. Arnold Goldburg, the scientist who had been handling the SST environmental issue for The Boeing Company, reported that he was called in by a company vice president who wanted to know: "What's this ozone problem you've been telling me about for three years? It doesn't exist!"

The whole thing was undeniably a mess. How it happened is fairly easy to pin down; it happened because of the way in which the results of the CIAP study were released to the public. The complete study itself was written by the CIAP scientists and was full of scientific data, calculations, charts, and graphs taking up some 7,200 pages. This was condensed to a "Report of Findings," a document of 850 pages. The "Report of Findings" was further reduced to a slim, 27-page document called the executive summary. This summary was written by Alan Grobecker and by Robert Cannon, then assistant secretary of the Department of Transportation. The substance of the summary reached the public in the form of half a dozen paragraphs of newspaper copy, for those who chose to read even that much, or, more likely, in the form of a single headline reading: "SST is cleared on ozone." Somewhere between the 7,200 pages and that single headline, the whole thing went off the tracks.

Why this happened is another question—one that launched the CIAP scientists into yet another round of bitter recriminations, emotional turmoil, and political controversy that would continue to fester throughout 1975 and would ultimately end in a congressional investigation into the possibility that the Department of Transportation had been influenced by political pressure. It was a frus-

trating and debilitating fight for the scientists, one that was waged
to regain lost ground that should not have been lost and therefore a
battle that should not have been necessary.

The trouble started with the executive summary, which was publicly released on January 21, 1975, several weeks before the full
"Report of Findings" was widely released and many months before
the complete CIAP study was generally available. The summary
was intended for all those people—congressmen, the media, other
nonscientists—who would never read the whole report. Its summa-
tion of the scientific findings was therefore crucial, and this is why
a furor erupted over the wording of the section entitled "Principal
Scientific Conclusions." Nowhere does the summary explicitly state
that the concerns raised in 1971 about large fleets of advanced SSTs
—concerns that had prompted the study in the first place and
presumably required some sort of answer—were fully justified. Such
a statement had in fact appeared in early drafts of the summary,
but it had been quietly dropped somewhere along the way.

Instead, the summary chose to emphasize, in its first principal
scientific conclusion, that the present-day SSTs and those scheduled
to go into service—some thirty Concordes and TU-144s in all—
would cause *climatic effects* too small to even be detected (emphasis added).

The second conclusion said that future harmful environmental
effects (which are not specified) can be avoided if "proper measures" are taken to develop nonpolluting engines and fuels. It is only
the third conclusion that makes even oblique reference to the problem of future fleets: "If stratospheric vehicles beyond the year 1980
were to increase at a high rate, improvements over 1974 propulsion
technology would be necessary to assure that emissions in the stratosphere would not cause a significant disturbance of the environment."

The words "ozone," "ultraviolet radiation," "biological effects,"
and "skin cancer" did not appear anywhere in the principal
scientific conclusions, nor did reference to the defunct Boeing SST,
which had originally precipitated the CIAP study. For what was
supposed to be a forward-looking exercise, the summary also dis-
played a puzzling avoidance of specifics about the potential effects

of future SST fleets. Projections of fleet sizes are not given. There are no clear and concise calculations of what future fleets of various sizes would do to the ozone layer. The discussion of the relationship between skin cancer and fleet size is extremely confused, and no figures are given. The distinction between the present-generation SSTs like the Concorde and the TU-144, and the more advanced kinds (sometimes called large SSTs) is blurred; yet the latter represented many times the environmental threat of the former. All of this information is in the body of the CIAP report but is nowhere to be found in the executive summary—at least not in any readily understandable form. *Science* magazine would later write that the summary "can only be described as misleading, whether deliberately or not."

But it is not entirely fair to blame the summary, despite its short-comings, for the newspaper headlines. The offending second paragraph of the AP story (saying that a fleet of advanced SSTs would not have weakened the ozone shield) resulted not from the summary but from the press conference at which it was released. In his own defense, Grobecker would later claim that he was misquoted. A look at the transcript of the press conference reveals that it is more precise to say that he was clearly misunderstood, and for that he is not entirely without blame. The exchange between Grobecker and one of his questioners, characterized by an excess of verbiage and a minimum of communication, is a classic illustration of the pitfalls of press conferences.

QUESTION: Dr. Grobecker, how would this study have changed if the United States had continued development of the SST?

GROBECKER: I think that's a sort of a question which implies the sense of—is this study directed by political considerations?

QUESTION: No, just asking you for your factual analysis of how the study would have been impacted if the United States had not turned back.

GROBECKER: Well, I don't think it would have been impacted at all. I think there has not been the slightest degree on the conduct of the study of political pressure. To any directed answer, and I think that by this time there might not have been enough airplanes flying to produce a measurable effect, so I find it difficult to see any way in which the outcome, the answer, could have been affected.

Grobecker did *not* say, as the AP story suggested, that a future U.S. fleet of SSTs would not have harmed the ozone layer. What he clearly meant to say was this: Even if the United States had proceeded with its SST program as planned, there would not have been enough of the planes flying *by the time the CIAP report came out in early 1975* to have made any noticeable impact on the ozone layer—or, therefore, any noticeable impact on the CIAP findings. (Had the U.S. SST program survived, the U.S. "fleet" would have consisted of little more than two prototype planes by 1975.) The point Grobecker was making at the press conference was essentially the same one he had made in the executive summary concerning the Concordes—there just weren't enough of them yet to make a difference. He was also clearly anxious to dispel any notion that a political decision in favor of the U.S. SST program might have succeeded in prejudicing the CIAP results.

In defending Grobecker later, Cannon would express difficulty in understanding how the statement could have been interpreted to mean that a full fleet of U.S. SSTs would not affect the ozone layer. But this is not difficult to understand at all. Grobecker's answer not only lacked clarity, it also was not responsive to the sense of the question he was being asked. What the questioner meant (although he did not make himself clear either) is: What effect would a full-sized fleet have produced? What Grobecker should have responded is that if the projected commercial fleet of SSTs had materialized—if, in short, the U.S. program had been a success—then, yes, it would have posed environmental problems.

It's hard to imagine that an answer like Grobecker's would pass, but it is in the nature of press conferences, which operate largely on a blunderbuss technique, that such things do get by all too often. No one at the conference seemed inclined to backtrack for clarification of the statement.

In light of the turmoil that the SST had continued to cause, off and on, for more than five years, it is surprising that many of the major newspapers did not choose to staff this press conference with their own specialist reporters. The fact remains that they didn't, and the AP story was the one that received the widest circulation. It appeared in several major papers, including the New York *Times* and the Washington *Post,* and the tone it gave to national news coverage of the event set the stage for the fallout of censure that descended on the CIAP scientists. It was also responsible for the headlines that essentially gave a blanket pardon to all SSTs— the result of the persistent lack of distinction between existing SSTs and future SSTs. Interestingly enough, British reporters did not make the same mistake, perhaps because Britain's involvement in Concorde made them keenly aware of such distinctions. The country's increasingly desperate efforts to keep Concorde in production and financially solvent made the gap between the existing fleet and a large commercial fleet painfully obvious.

The British scientific journal *Nature* was one of the few to explicitly acknowledge that, the executive summary notwithstanding, the three-year CIAP study had *not* banished fears for the ozone shield and that it had *not* refuted the scientific theories on which those fears were based. "The argument about the effect of SSTs on the environment is therefore unlikely to be stilled by this report," *Nature* concluded. It turned out to be a prophetic statement, as events a few days later at a CIAP meeting in Boston proved.

The meeting started on February 4, about two weeks after the release of the executive summary. The CIAP scientists had those two weeks to absorb the denunciations being handed out by the press, and they were collectively in a black and mutinous mood. The muttering began in the hallways on the first morning before the meeting even started and, when it started, so did the fireworks.

Grobecker was first on the agenda, and he read a paper based on the executive summary that had been featured at the ill-fated Washington press conference. His presentation further irked many of his listeners, for he managed to convey the impression that there was, in effect, no SST problem at all—that the present fleet represented no hazard whatsoever and that taking care of future fleets by cleaning up the engines was a straightforward, even simple task, if not necessarily a popular or inexpensive one. Indeed, there was in the executive summary and in Grobecker's pitch a kind of dogged persistence that the necessary engine improvements could be made in time, that they could keep pace with the growth of SST fleets, that they could prevent these newest of the flying machines from spewing nitrogen oxides all over everywhere.

The summary, for example, estimated that the then projected NO_x emissions could be reduced by at least a factor of six by redesigning the engines using existing technology.[2] It said that such a reduction, which would take ten to fifteen years to implement at a cost of about $50 million in development money for each engine type, would allow for some expansion of Concorde-type SST fleets. The summary also stated that achieving a sixty-fold reduction in NO_x emissions would take perhaps another ten years, allowing for even further growth in SST fleet size. In fact, Grobecker concluded that for twenty to twenty-five years, stratospheric air travel could expand to meet demand without adversely affecting the environment.

Grobecker's technological optimism soon became a major sore point among his scientific colleagues, particularly since it communicated itself so strongly to nonscientists. For example, one newspaper article stated that the ability to develop low-emission engines and fuels at a reasonable cost was "well-advanced and attainable." In fairness, Grobecker did not put it quite like that himself, but the statement is not inconsistent with the tone he set in the executive summary.

Grobecker's presentation in Boston before his scientific colleagues was somewhat more tentative. Rather than saying that current technology could realize a sixfold reduction in NO_x emissions, he said that this was a *goal* for 1985 to 1990, which is not at all the same thing. He acknowledged that, while NASA and the engine manufacturers were trying hard, the technology for this sixfold reduction had not yet been demonstrated, and he added: "It is a far cry from

2 According to one source, this factor of six had a rather curious origin. Grobecker apparently adopted it after being accosted at a cocktail party by an engineer for one of the aircraft engine manufacturers who reportedly complained that "you guys are always talking about reductions of a factor of one hundred. That's ridiculous. Now, if you were talking about a factor of six, I might not have a heart attack."

demonstration in the laboratory to having an engine which has been tested and certified for flight operations."

Grobecker further noted that the sixtyfold reduction listed as the projected mimimum for NO_x emissions was in fact "a theoretical limit"—something that was not spelled out in the executive summary.

But the tone of the paper he gave in Boston was still basically optimistic, and Grobecker's faith in the technological fix provoked a palpable skepticism in the audience. The CIAP scientists felt he was greatly overselling the ability of engineering and technology to control future events, but it was the spokesmen for the aviation industry who really undermined his position.

During the discussion after Grobecker's speech, Arthur Nelson of Pratt & Whitney Aircraft got up to say: "I would like to correct one misunderstanding: that the aviation industry could achieve a sixfold NO_x reduction with present technology." Nelson said they were unable to achieve even a twofold reduction in a practical engine, despite intensive efforts. "A sixfold reduction will require, unfortunately, new inventions or new ideas which we do not have today." He concluded by saying that even if such an engine could be developed—something that was not yet established—it would require a long and expensive research and development program.

This theme was backed by Donald Bahr of General Electric's Aircraft Engine Group, who said in his talk that technology significantly different and more complex than that which existed would be needed to produce engines with extremely low NO_x emissions.

A NASA engine expert was quoted in the press at the time as saying that reductions of a factor of sixty might be "approaching the realm of impracticality."

Several weeks later, the National Academy released a study that said that a factor of ten to twenty reduction in NO_x emissions might well be obtained within ten to fifteen years, *provided that* adequate resources and a high priority in funding were given to such a program (emphasis added).

During a panel discussion following Grobecker's speech, Robert Cannon warned that squabbling among scientists (he used the term "perceived dichotomy") would do nothing to encourage the government to finance the measures needed to prevent the SST from having an intolerable environmental impact. Cannon, who had by then left the Department of Transportation for a position at the California Institute of Technology, was insistent on the matter of preventative measures, which he saw as the answer to the SST problem in preference to abandoning the plane. Engine research, atmospheric studies, regulatory activities—all these were going to cost a lot of money, he told the delegates in Boston. If scientists appeared to disagree with each other, the government would "simply conclude that it is not clear that anything should be done and they will not fund the needed preventative measures."

Cannon emphasized the term "perceived dichotomy," seeming to imply that there were no substantive disagreements among the scientists. This may have been largely true with respect to the actual scientific data contained in the full CIAP study, for there was a reasonable consensus on that. But it was not true with respect to the executive summary, and if Cannon thought his warning would quell the rising dissension, he was badly mistaken. The scientists in Boston were in a querulous mood, and they were not about to be put off with the suggestion that the dispute about the executive summary was merely a trifling exercise in semantic quibbling. Its impact on the public and its reflection on their own credibility had been too great for that. The resentment that had been building up in the two weeks since the release of the executive summary let loose at the Boston meeting. The scientists were stung by the injustice of an attitude that suggested there was no SST problem. After all, if there was indeed "no SST problem" this was solely because there were practically no SSTs. But the scientists had not dreamed up the fleet of hundreds of advanced SSTs on which they'd based their calculations, predictions, and warnings; those figures came from the Department of Transportation and from the aviation industry.[3] The fact that the fleets had not yet materialized—and maybe never would—did not make the scientific theories about

[3] In 1971, both the DOT and the aviation industry were talking about 500 Boeing SSTs as an economically viable fleet. In July 1974, the Environmental Protection Agency projected 375 Concordes for 1990. At the Boston meeting, CIAP projections of the "most likely" fleet indicated there would be 241 large SSTs by 1990 and 615 by the year 2000.

their environmental impact wrong. The CIAP report could in no sense be construed as a green light for futher large-scale SST development.

One of the things that particularly irritated Hal Johnston was the fact that Grobecker seemed to use one set of figures when he was talking about the economic prospects for the SST and another set when he was talking about the environmental impact. In a January 1974 paper, Grobecker had projected a market for 500 Concorde-type SSTs in 1990, 1,400 Concorde and large SSTs by the year 2000, and a staggering 5,000 or more large SSTs by the year 2025. It seemed quite wrong to Johnston that Grobecker should make market predictions of 500 to 5,000 SSTs while limiting the principal conclusions of his environmental impact statement to 30 Concordes and Tu-144s or limiting his remarks at the Washington press conference to the environmental impact of 2 prototypes.

The CIAP scientists gathered in Boston were perhaps most outraged by the attacks on their credibility that had resulted from the fiasco over the executive summary. They were still smarting from the potshots being taken by outsiders who had read of the executive summary's "no problem" stance and carried it one step farther, saying that if there was no problem then these scientists—these troublemakers—were guilty of crying wolf and of all the multitudinous sins that follow from this, such as wrecking the economy, destroying jobs, and undermining the U.S. aerospace industry.

That Hal Johnston should be the most vocal of the critics was hardly surprising. Perhaps more than anybody else there, he had a personal stake in the outcome of the debate. He had been one of the first scientists into the SST fray and was certainly among the most prominent. Since his high public profile had made him a vulnerable target for those in the opposing camp, he could arguably treat any assault on the credibility of scientists in this matter as something in the nature of a personal attack. Granted, he had voluntarily adopted that high profile, but that did not mean that he would quietly suffer what he considered to be unfounded criticism. At the time of the Boston meeting, he was entering his fifth year of a deep, personal, almost consuming involvement in the imbroglio. It had not been an easy four years; more than once he had been compelled to defend his research, his calculations, his point of view against fierce assault. But now the verdict was in; the CIAP study would vindicate his concern. He was not about to let the misguided publicity over the executive summary snatch that away. He cornered Grobecker at a gathering the night before the meeting started and told him he thought the executive summary was misleading. He had written his own version, based on the full CIAP report, and the next day at the meeting, he got up to read its conclusions.

Johnston started by saying that Grobecker's executive summary, though consistent with the CIAP findings, "suggests to me an adversary position in support of the SST. Some questions which could have been simply and clearly answered were not answered. Thus it seemed to me that somebody should write an adversary set of conclusions consistent with the CIAP findings—that is, anti-SST—for the sake of balance, since the other was not neutral. Now, I've never considered it my job to be an anti-SST advocate, but late last week, it seemed somebody should do the job for this one time." In other words, he was saying that, despite Cannon's admonition, if there had to be a "perceived dichotomy," so be it.

Johnston first stated that 125 Concordes and TU-144s would decrease ozone in the Northern Hemisphere by .5 per cent and increase skin cancer in the United States by 5,000 cases per year or 100,000 over the 20-year lifetime of the aircraft. The 375 Concordes projected by the Environmental Protection Agency for 1990 would have threefold greater effects.

Johnston's second conclusion was that the 500 Boeing-type SSTs talked about in 1971 would decrease ozone 12 per cent, increasing U.S. skin-cancer cases by nearly 120,000 a year, or 2.4 million over 20 years.

The third conclusion dealt with Grobecker's projected year 2025 fleet of 5,000 large SSTs. These, Johnston said, would reduce ozone by a factor of three or more. Even if engine emissions were improved sixtyfold—a theoretical limit—ozone in the Northern Hemisphere would drop 5 per cent. "It appears improbable that very large fleets of SSTs can be made safe even by reduction of the NO_x emission index," Johnston concluded.

Johnston was particularly irritated by the fact that the summary's first scientific conclusion made reference to "climatic effects" rather than health effects—that is, skin cancer. "Climatic effects are one thing and health effects are another," he later wrote in a letter to *Chemical and Engineering News.* "Nobody ever implied that 30 Concordes would change the climate, and it is trivial to assert that they won't. . . . The principal conclusions never identified the health threat and this omission was deceptive. . . . The DOT in its report of findings plays the role of an advocate of the SST industry."

Both Grobecker and Cannon would later argue that they considered the phrase "climatic effects" to be a generic term that embraced the ultraviolet effect. But this is not an interpretation shared by the majority of other scientists, who rarely, if ever, lump the two together. It is true that Grobecker inherited the emphasis on climate from the early days of the SST controversy, but the CIAP study itself had been totally dominated by the ozone question, and the full report clearly distinguished between climatic and ultraviolet effects. In fact, Grobecker himself makes this distinction in the body of the executive summary. Moreover, early in the CIAP program, when Cannon announced that Grobecker had been chosen to head it, Cannon said the purpose of the study was to determine whether high-flying aircraft "could have any significant effect on the climate *or on people's health.*"

However, at least one of Johnston's objections to the way the summary handled the skin-cancer question was patently unfair. The summary talks about the skin-cancer problem involving "an assumption, supported by some scientific evidence but not proven by experiments on humans." In a letter to the New York *Times* (which was not published), Johnston rather indignantly claimed that "this statement shows the fantastic degree of proof DOT requires before it accepts any conclusions adverse to the SST—cancer experimentation on humans." In fairness, the summary does not suggest that such experiments should be required—merely that such proof does not exist.

Johnston's main message—a theme he would return to often—was contained in an exhortation to his colleagues to read the "fine print" of the CIAP report of findings, not just the summary's conclusions or the headlines. He himself had done a thorough, sentence-by-sentence analysis of the executive summary. (His an-

noyance with it is reflected in some of the marginal comments
scribbled on one copy: "Very dubious statement." "Trivial and misleading." "Backhanded way to admit the significance of the problem." "Tricky!") But if he really expected others to do such an exhaustive analysis, or believed that this could counteract the
damaging effects of the executive summary and the newspaper stories, he was being terribly naïve. The summary is as far as most people got or were ever likely to get. As one of Johnston's colleagues
later remarked, it was unlikely that *anyone* (with the possible exception of Hal Johnston) had ever read the entire CIAP report, including the scientists who worked on the study. "Certainly no public official did or will. So we're left with summaries and newspaper
articles."

After Johnston sat down, Fred Kaufman took the floor. Kaufman, a physical chemist from the University of Pittsburgh, is in
many ways the antithesis of Johnston. Kaufman is a large, heavy-set
man with a booming voice that still carries the accent of his native
Austria. Kaufman is known for his humorous erudition, and he has
a talent for enlivening otherwise dull scientific meetings with his
barbed wit (although, in recent years, he has coupled this with a
tendency to make rather personal attacks on individuals in public).
In Boston, he began by saying: "I hope my great friendships with
Drs. Grobecker and Cannon will not suffer from the blast I am
about to deliver. . . .

"I find it supremely ironic," he went on, "that a program that
looks thirty years ahead does so poorly at looking one day ahead:
from the day the 'Report of Findings' was released to the next day,
when it was reported in the news media. The impression gained by
the educated news-follower was that there is no ozone problem. I,
of course, agree with Hal Johnston that the 'Report' did a terrible
public-relations job and that you are suffering the consequences.
Though most people in this room know what CIAP has accomplished, I was wishing, until Hal came along, that someone would
set the record straight the way he has."

As proof of the damage done, Kaufman brandished an editorial
from the Pittsburgh *Press*, an evening newspaper serving a metro-
politan area of 2.4 million people. On the matter of SSTs and the

ozone layer, this editorial said, the scientists had those several years before spoken "unscientific nonsense."

"The phony ozone argument has no place in rational scientific discourse and no place in the SST debate. It is useful to know the facts now, even if exculpatory evidence that arrives after the hanging is of small comfort to the deceased."

Kaufman also questioned the summary's "sudden obsession" with the concept of minimum discernible change in ozone (something Grobecker's own CIAP staff had already privately objected to). This referred to the smallest change in the ozone layer that could be detected by a monitoring system and was rather arbitrarily defined in the summary. In using this concept, Grobecker was making an assumption that, as he himself once put it, "any change smaller than the smallest effect which could be felt or sensed would be tolerable." This was of course absurd. It implied that the environmental impact was unimportant as long as we could not detect it. But the ability to detect an effect depends simply on how good the technology is; it has little to do with the intrinsic seriousness of the environmental problem. It's like arguing that just because you don't have an instrument to detect cyanide with great precision, it's all right to drink it in small doses.

It was this concept that Kaufman challenged at the Boston meeting. "Suppose the minimum discernible change in ozone had been 20 per cent. Would that give us license to change it by that much?"

Despite the complaints, Grobecker remained his affable self. Alan Grobecker is in many ways the basic mild-mannered scientist. He is bespectacled, soft-spoken, and unfailingly courteous. He does not lose his temper when you ask him about the Great Executive-summary Snafu—at least not outwardly, although sometimes, when you catch the trace of exasperation that creeps into his voice, you can see he would like to. He did not lose his temper in Boston, yet he could not have welcomed or enjoyed the conundrum he found waiting for him there.

"Professor Johnston always has a challenging thing to say," he began, adding that he did not disagree with the figures Johnston had cited. "That is a perfectly satisfactory way of describing things, although"—a slight testiness—"I don't choose to impute unfavorable actions to others."

Grobecker regarded the matter as more of a semantic fuss than anything else. "As can often be the case with such a complicated subject, the data on which all agree may still lead to conclusions or recommendations with which not all agree. . . . The problem is a little like answering the question of whether a cup is half empty or half full. The answer can be a matter of taste."

There was also the fact that Grobecker was to a certain extent the victim of the media's penchant for oversimplifying complex scientific issues. Cannon described the problem well: "One is asked the disarmingly simple question, 'Is there a problem?' If one answers by saying, 'There is no serious problem if we take suitable preventative measures,' the resulting headline reads 'There Is No Problem.' If one answers by saying, 'There is a serious problem unless we take suitable preventative measures,' the resulting headline reads, 'There Is a Serious Problem.'"

Grobecker's protest that it was all the media's fault was by no means the worst defense he could have offered his colleagues. At such a gathering as this, it is not a difficult task to make the media the villains of the piece. As a rule, scientists do not much love the media, and they are entirely predisposed to believe absolutely the worst of reporters. Enough of the scientists in Grobecker's audience have their own horror stories about the failings of the media—and enough of those stories are true—that his complaints about the media could normally be expected to arouse much sympathy.

Nevertheless, Grobecker's plaintive cry of "misquote" did not bail him out. There was a certain unforgiving cynicism at work here—a feeling that, damnit, Grobecker should have known better than to let this happen. He had been handling the media for more than three years now; surely he was not so naïve as to think that they would read the fine print, that they would wade through the charts and graphs, to piece it all together?

To appease the critics, Grobecker called another press conference to set the matter straight. (Cannon says they had decided to do this immediately after the January 22 press stories.) He considered this to be a prompt response to the misleading news reports about the executive summary and he would later write, in a letter to *Science* magazine, that the correction got "wide coverage in the press and on television." But it clearly did not—corrections rarely do—and it did not really make anyone happy, for in the end they were saddled with that "perceived dichotomy" after all.

"Experts Differ on Ozone Study" read the headline in the New York *Times*. "Possible Effect of SST on Stratosphere at Issue." The story, written by the *Times'* science writer, Walter Sullivan, began: "Sharp differences as to the significance of a four-year study of potential supersonic transport effects on the stratosphere . . . erupted here today. . . ."

The subject of the executive summary went into a state of abeyance at the Boston meeting. The bellicose mood persisted and the grumbling continued during coffee breaks and over cocktails, but there was the real business of science to attend to, and so the researchers turned to that.

But if Grobecker thought he had seen the end of it, he was mistaken. The opening shot in the second round came from Thomas Donahue, an atmospheric scientist from the University of Michigan. Donahue had not been directly involved in the CIAP study but, as an official of the American Geophysical Union, he had agreed to sanction, in the name of that body, a scientific review of drafts of the CIAP report. He had cause to regret this decision when he saw the executive summary, for he feared that his participation in the review process made it look as though the AGU had also endorsed the summary. This was emphatically not the case. Donahue was as annoyed as Johnston at the evasiveness of the document and as outraged by the news reports that followed it. In fact, Donahue immediately called Grobecker to insist that the CIAP manager correct the impression left by the news stories. "I made two phone calls to him and both times I was rebuffed," Donahue said. According to Donahue, Grobecker said he had checked with his superiors and had been forbidden to issue the disclaimer of the news stories that Donahue was demanding he make. Grobecker denies this. "Nobody said word one to me." As far as he was concerned, the AP story had published an "astonishing misquote," but Grobecker stood by what he had said in the summary. "I was not going to be stampeded into making a disclaimer of what my report had said." Donahue warned Grobecker that he would not hesitate to criticize the summary if given the chance (and he later did just

that, before congressional committees on two separate occasions).
Both Donahue and Johnston then tried to have letters published in
several newspapers, but without any very great success. Finally, at
the end of March, Donahue had a letter published in *Science* magazine and in *EOS*, a journal of the American Geophysical Union.
In it, he offered his own interpretation of what the summary's conclusions should have been. If five hundred Boeing SSTs had been
built and flown as planned, they would probably have reduced the
ozone layer by 10 to 20 per cent. The fleet of five thousand large
SSTs would have reduced the ozone layer by two thirds or more.
The engines would have had to be sixty times cleaner than planned
to keep the ozone depletion to a level of 5 per cent.

These conclusions, Donahue caustically observed, could be determined by any perceptive person with good deductive abilities and
the patience to root through footnotes to tables and other assorted
fine print in the executive summary.

Donahue concluded the letter by saying that Grobecker had twice
refused his appeals to correct the impression created by the AP story
and, in consequence, "those who raised the alarm have been effectively discredited and stand accused of providing damaging counsel
to this country. I hope this letter will repair a little bit of the damage that has been done."

Grobecker responded with a short note published in the same
issue of *Science*. He pointed out that he had released a correction at
the Boston meeting, then proceeded to give an explanation of how
to figure out the effects of a fleet of Boeing-type SSTs. This explanation proved Donahue's essential point—that it was indeed necessary to go rooting around in the tables of the summary to calculate
the effects of the Boeing SSTs. Grobecker noted in his letter that
"the impact of five hundred Boeing SSTs is *implicit* in a comparison presented in Table 15 of the 'Report of Findings'" (emphasis
added), but he does not actually do the calculation, nor are we told
why such a central calculation is only implicit in a report intended
to answer such questions explicitly.

The reverberations of the executive-summary controversy continued into the spring of 1975. On April 1, the National Academy of
Sciences released its review of the CIAP study confirming the SST
threat to the ozone layer. The news headlines this time were very

different. "Jet study warns of skin cancer risk," the New York *Times* announced.

At a press conference, Henry Booker, chairman of the Academy committee that prepared the report, took pains to again debunk the news stories that had followed the CIAP summary. He released a letter he had received from Robert Cannon that stated that the CIAP study "clearly supports the validity of concerns voiced by perceptive scientists in 1970, that strict measures might be needed to protect the environment from the effects of future large-scale flight operations in the stratosphere."

This was not the only letter Cannon wrote in his attempts to clear up the misconceptions that arose from the executive summary. He sent one to the *Christian Science Monitor,* which had printed the Roscoe Drummond column charging that the SST had been killed on the basis of "ecological myths."

"It is not true that the concerns voiced in 1970 are mythical," Cannon wrote. " . . . As the report makes clear, the concerns raised in 1970 about ozone are valid concerns."

Cannon was still writing letters like this a year after the summary was released. One appeared in *Science* magazine in January 1976, just before Transportation Secretary William Coleman was to decide whether or not to let the Concorde into the United States for a trial period. All of Cannon's letters followed a similar pattern. Like Grobecker, Cannon was a technological optimist. He believed that it was technically and economically possible to develop clean engines, but research had to start on them at least ten years before they were needed. He stressed the urgency of beginning work on the necessary preventative measures right away. Neither Grobecker nor Cannon denied that a problem existed; they always explicitly acknowledged that large-scale SST operations did pose a threat to the ozone layer. It was just that they did not share the philosophy of others that technology might not be equal to the challenge of solving the problem. And they certainly did not subscribe to the view—which is a value judgment, after all—that we would be better off without SSTs anyway. Scrapping SSTs was not, to their minds, a palatable option.

Cannon was deeply disturbed by the controversial and highly political aftermath of the CIAP study, dismayed that the project's accomplishments were being buried by the executive summary controversy. "I've lost an enormous amount of sleep over it since," he said.

For Alan Grobecker, 1975 was not a very good year either. "Sure I took it personally," he says of the controversy, a slight edge of bitterness creeping into his voice. "It ruined my year. I don't like being charged with a lack of integrity or with distorting the facts."

In retrospect, he can be philosophical about the uproar, though there are times when he wishes he had followed a favorite maxim: "The prudent man seizes every opportunity to keep his mouth shut." He can even laugh at the suggestion that he should have been more sophisticated in his handling of the press, but it is a short, sharp laugh, almost a snort, which conveys a genuine surprise that people actually expected this of him. He did have a chat with the reporter who wrote the AP story, but did not feel compelled to pin him to the wall to extract a *mea culpa*. "That's not my style," says Alan Grobecker. Mildly.

Counteracting the misunderstanding about the CIAP study took considerable time and effort on the part of several CIAP scientists—effort that could have been productively directed elsewhere. But the effort was necessary for two reasons, the first involving the emerging spray-can controversy; the second involving the political battle shaping up over landing rights for the Concorde in the United States.

By the time the CIAP executive summary was released in January of 1975, the Spray-can War was in full swing, and it was giving every evidence of becoming a worse dilemma than the SST ever had been, for the fluorocarbon problem was here now, while the SST problem was not. The advanced SST fleets, after all, did not exist, were not making large profits for their builders and operators. That they might someday exist was a possibility, but it was not a reality. Fluorocarbons, on the other hand, were a reality—and the scientists could see that they would soon be called upon to give advice on what should be done about this viable, billion-dollar industry. Never had their need for caution and public credibility been higher. This is why they were so upset by the repercussions of the CIAP executive summary. As Donahue put it in his letter to *Science*:

"The principal public result of this report so far has been to cast doubt on the serious nature of the questions now being raised by atmospheric scientists concerning threats to the stratosphere by (man-made) pollutants other than nitrogen-oxide emissions from SSTs."

There is no small amount of irony in this. The CIAP experience has given scientists the knowledge and the tools they needed to understand and cope with the fluorocarbon issue with an efficiency and thoroughness that would not have been possible just three years before. But had it, at the same time, damaged the credibility that is their only currency in the social and political wars?

And there was another concern, perhaps even more dismaying: The CIAP scientists became increasingly suspicious that political pressure may have played a role in the timing and wording of the executive summary. If so, it meant that politics had indirectly been responsible for distorting scientific information, or at least the manner in which that information reached the public. What triggered these suspicions was the protracted battle over U.S. landing rights for the Concorde.

CHAPTER FIVE

You Are Cleared to Land

They came.
They saw.
They Concorde.

—Letter to *Time* magazine, March 1, 1976.

For an event that was supposed to mean life or death to one of the largest technological ventures of modern times, it had all the trappings of an antic situation comedy.

There was William Coleman, the rotund Secretary of Transportation, closeted in the phone booth, the door barricaded by a military aide (from whom he had to borrow the fifteen cents for the call, which he then misdialed), telling President Gerald Ford of his decision to allow the French-British Concorde supersonic transport into the United States for a trial period of sixteen months.

There was an amusing and obvious symbolism to this little tableau. Coleman made the call while hustling to his Concorde press conference like a portly, waistcoated Rabbit en route to Wonderland, and it had the effect of demonstrating that the White House had kept hands off the politically sensitive decision. In short, the President was getting a beat of only a few minutes on the rest of the world.

Meanwhile, over at the Department of Transportation (DOT) headquarters, all was in chaos. Dozens of reporters, French and British embassy and airline representatives, environmentalists, and assorted hangers-on were crowding into the large press-conference room, tripping over television cables and each other and cursing at the photographers and TV crews who were, as usual, making a nuisance of themselves.

An hour before, at 11 A.M., about fifty reporters had been herded into a tiny, narrow room to read advance copies of Coleman's decision. Outside, half a dozen burly, uniformed guards stood watch at the door, evidently determined to thwart any attempt by the press to leak the news. All present presumed that their holsters did not conceal paper punchers. Veteran Washington reporters could not recall anything quite like this having happened before, but they seemed more amused than angry about it.

Inside, the DOT public-relations chief, Ramon Greenwood, started into his spiel. As his aides hovered protectively over the stack of documents, Greenwood announced that he wanted to go over the "ground rules," which he said had been arrived at after long discussions within the DOT and with representatives of the media. Since the subject at hand was "extraordinarily sensitive," he wanted all the reporters present to promise that, on leaving the room at noon, they would go immediately to the press conference and would not write or release their stories until after it was over. He added that representatives of the British and French governments in Washington, London, and Paris would also receive the document under the same restrictions.

Greenwood started off smoothly enough, reading from a prepared text, but he ran into flack almost immediately from the wire-service reporters, whose deadlines are always half an hour ago and who are extraordinarily sensitive to an advantage of even a few minutes. For one thing, no one had consulted *them* about these ground rules, and for another, the wide distribution of the document invited abuse. There were dozens of ways the news could get out before the end of the press conference. With the potential for leakage so high, there was no way they were going to sit idly around in the press conference twiddling their thumbs for an hour. The man from the British service, Reuters, popped up to say he did not "propose to see Reuters slaughtered on this particular story." He noted that it was customary in such circumstances for the wire services to file their stories on a hold-for-release basis so they wouldn't get caught flat-footed if there was a leak. The French and U.S. services shouted their agreement. There ensued what the bored daily newspapers reporters, whose deadlines were a comfortable few hours off, later agreed was fifteen minutes of acrimonious debate, which was abruptly resolved by the Reuters man, who announced in polite, but clipped, British tones that he was leaving as soon as the door was opened and implied that he didn't care if the guards shot at him or not (they didn't). Nonplused, Greenwood mumbled something about putting everyone on their honor and succumbed to the mounting clamor to get on with the show.

The cause of all this emotion was a sixty-one page document, released on February 5, 1976, in which Coleman outlined his reasons for allowing Air France and British Airways to operate six Concorde flights daily into Washington's Dulles International Airport and New York's John F. Kennedy International Airport. He ordered that a monitoring system be set up at both airports to measure the noise and pollution from the aircraft and stipulated that the Concorde flights had to originate from Paris and London; French and British citizens were not to be exempt from the noise to which U.S. citizens would be subjected. The sixteen-month trial period was to allow a year of Concorde operations, plus four months to analyze the data from the monitoring system. But the whole thing could be called off on four months' notice—or immediately, in case of an emergency deemed harmful to the health or safety of U.S. citizens.

This was widely regarded as a make-or-break decision for the Concorde, which badly needed the lucrative North Atlantic run. Predictably, reaction to it ran the gamut. French officials grumbled that the Concorde deserved a "true yes," not a qualified one. Opponents called it a fraud and a sellout, and moved swiftly, but without success, to overturn the decision in Congress and the courts. In fact, the Coleman decision touched off a complex legal battle that rapidly gave promise of entangling all interested parties for some considerable time to come. As usual, the Concorde was not just a plane, but also a political happening.

It cannot be said that the fight over Concorde landing rights revolved primarily around the ozone issue. The main environmental issue was the noise levels, particularly on landing and takeoff, and the political fight against Concorde landings was spearheaded largely by congressmen whose constituents would be subjected to those noise levels. Concorde opponents were also greatly exercised about the possibility that politics and foreign policy had played a strong role in assuring that landing rights would be granted. There were persistent rumors that the Nixon White House and the State Department were mixed up in this, that they had been leaning on the Department of Transportation and the Environmental Protection Agency while making secret assurances to the French and British that all would be well.

For the scientific community, the most worrying possibility was that the scientific information gathered during the CIAP program —or at least the timing and manner of its presentation to the public —had been perverted for political ends. Some saw, in the trouble-fraught quest for U.S. landing rights, the obvious rationale for the CIAP summary's curious preoccupation with the Concorde. They were alarmed, as Tom Donahue put it, that the summary had been strongly influenced by political pressure "to prevent the CIAP report from creating a climate in which it would not be possible to grant a license to land the Concorde" in the United States (a charge persistently and vehemently denied by the authors of the summary).

Indeed, barely more than a month after the summary was released, with the misconception that there was "no SST problem" still lingering in the public mind, both the Environmental Protection Agency (EPA) and the Federal Aviation Administration (FAA) issued statements basically favorable to the Concorde. The FAA, in its draft environmental-impact statement,[1] concluded that the environmental consequences of a limited number of Concorde flights would not be so severe as to compel refusal of landing rights. Since this statement dealt with the ozone issue simply by appropriating the conclusions of the CIAP executive summary (including the first one, emphasizing the minimal climatic effects of thirty Concordes), the complaints against the executive summary were soon visited upon the draft environmental-impact statement. A letter from the President's Council on Environmental Quality to the Transportation Department admonished the DOT to cite the full

[1] All technological developments now require impact statements. Moreover, a draft must be publicly released first, so interested parties can comment before the final impact statement is written.

text of the CIAP report, not the executive summary, in its final environmental-impact statement.

The use of the summary's controversial conclusions made it inevitable that Hal Johnston and Tom Donahue would become deeply embroiled in the political struggle over Concorde landing rights. On April 14 they both testified at a hearing on the FAA's draft environmental-impact statement, predicting that thirty Concordes would cause between twelve hundred and twenty-one hundred extra skin-cancer cases a year in the United States and perhaps ten to twenty deaths a year. Over their twenty-year lifetime, these Concordes would cause tens of thousands of cases a year.[2] Donahue told the FAA that it was being "flagrantly irresponsible," and he ended his testimony on an impassioned note. The CIAP study had upheld the warnings about the danger to life on earth posed by SSTs, and Donahue asked: "What was the use of all this study? Of what value is science and wisdom to this society if you deliberately disregard these clear warnings? What you propose to allow is much worse than not turning off already existing sources of pollution, for you are deliberately allowing a deadly new source to be born in the face of a clear, grave and official warning. . . . A decent regard for the future of mankind, only a tiny fraction of which wants to fly at two thousand miles an hour, demands that these aircraft not be allowed to begin to operate until it can be demonstrated that it will be safe to fly them in the numbers that economics will demand. To go the other way around is madness, like allowing a person to start using a little bit of heroin or smoking a few cigarettes a day on the grounds that either you will find a way to make a large habit safe or prevent him from increasing his use of the poison."

The FAA received approximately twenty-nine hundred comments on the Concorde, all but sixty of them opposed to the aircraft.

The release of the CIAP executive summary and the FAA draft environmental-impact statement touched off a fierce and acrimonious political battle in Congress. On January 31, Lester Wolff, a New York congressman who was one of the leading SST opponents,

[2] These numbers were consistent with the scientific knowledge at the time but, in light of more recent data, are no longer valid.

set in motion an investigation of the CIAP summary. Wolff and New York Congresswoman Bella Abzug also began relentlessly to pursue evidence that political pressure had been brought to bear on the EPA and the FAA. They were particularly interested in letters former President Nixon had apparently sent to the leaders of France and Britain assuring them that the Concorde would receive fair treatment. There was considerable speculation that Nixon had gone farther and had actually made commitments that the Concorde would receive favorable treatment. The exact nature of these commitments was not clear, but in one sense it hardly mattered; according to at least one report, the mere existence of the letters and the implication that they contained a commitment had been used to browbeat the U.S. federal regulatory agencies. Harry Pearson, environmental writer for *Newsday,* quoted Roger Strelow, the No. 3 man at the EPA, saying that in discussions with officials of the State Department and the National Security Council, "there was mention made . . . of a specific letter from Nixon to the heads of state in England and France. It can be construed as an assurance from Nixon to them that the plane would be allowed to land in this country."

The very idea of secret commitments outraged the Concorde opponents. For one thing, if Nixon had indeed attempted to make binding commitments, he had probably exceeded his constitutional authority. For another, such commitments made a mockery of the CIAP study, the environmental-impact statements, the public hearings, U.S. aviation laws and regulations, and congressional hearings on the Concorde issue. And so Wolff and Abzug began a persistent campaign to force the White House, then presided over by Gerald Ford, to release the Nixon letters. Their request was refused; the letters were considered part of Nixon's presidential papers and thus were tied up in the post-Watergate battle over ownership of the papers.

On July 24, 1975, the congressional investigation began. The purpose of the hearing was not to determine whether the Concorde should be allowed to land in the United States but whether the steps being taken by the FAA toward that end were correct, lawful, and carried out without prejudgment or undue outside influence. One of the first witnesses was Congressman Wolff.

Wolff charged that as far back as 1971, senior U. S. Government officials had begun making a series of "highly improper" commitments to the French and the British—commitments that not only came at crucial times in the decision-making process on Concorde but were also used in the United States "to bully officials in the executive branch into suppressing facts or making decisions which they otherwise might not have made."

According to Wolff, the most serious problem was the "apparent attempts by some in the executive branch to corrupt the actual scientific process itself . . . the DOT and FAA have systematically hidden, and in some cases apparently falsified, the actual physical and psychological effects of the Concorde SST on the earth's environment and its people."

In November, the congressional subcommittee took up the matter of the CIAP study. Among those called to give testimony on November 13 were Hal Johnston, Tom Donahue, Steve Wofsy of Harvard University, and Robert Cannon, one of the authors of the CIAP executive summary. Alan Grobecker, the other author, also attended the hearing, but he was not called to testify.

Charles Thone, a member of the subcommittee, reported on an interview he had done with Mike McElroy at Harvard. McElroy, he said, "stated emphatically that it was his opinion that the executive summary placed undue emphasis on those facts most favorable to the SST. He simply doesn't consider it an objective, balanced presentation. . . . Dr. McElroy believes the executive summary was overly influenced by Dr. Cannon and it does not represent the consensus of the study's participants. . . . He also questioned the propriety of releasing the executive summary so long before full 'Report of Findings.' This allowed the public to come to conclusions without the benefit of the further data, which might tend to support different conclusions."

Johnston and Donahue also reiterated their complaints about the executive summary. Johnston, whose written submission addressed itself to "some deceptive aspects" of the CIAP "Report of Findings," said that "proof that [the executive summary of] the report is deceptively worded is the fact that people were deceived." He cited the newspaper accounts that had "cleared" the SST and argued that, despite the corrections "many responsible people still do not understand the nature of the CIAP findings. They were more impressed by the initial press conference and its attendant publicity than by the later actual report or by the corrections."

Johnston said that the principal scientific conclusion should have been that "human activity can seriously reduce global stratospheric ozone." He argued that "Congress has a permanent responsibility in this new area and it should not be distracted from this responsibility by bland assurances that thirty Concordes will not change the climate."

Johnston did not explicitly charge that the CIAP executive summary had been subject to political influence, but he did quote Alexander Flax, head of the Institute of Defense Analyses (and a former boss of Grobecker's) advising the scientists to "carefully word it [the results of the study] not to seem to be pushing the NO_x problem as an effort to undercut and do-in the Concorde."

Johnston was questioned at length by the congressmen, some of whom seemed to flounder a bit in the scientific detail. (That scientific data were viewed with some trepidation was demonstrated by the remarks of subcommittee chairman William Randall of Missouri when Johnston took the witness stand. "You have . . . many graphs and charts. I know that you are a scientist. Do the best you can to help us understand what you are talking about." Later, Randall commented: "It is quite apparent that the Chair is probably worst of all as a scientist. We are going to try to stumble along here.") Randall asked Johnston about his calculations of skin-cancer cases. "You have given us a mathematical formula that goes clear across the page. And you have made the charge that these multiplications of the . . . 'disjointed' data in the so-called executive summary would be pretty difficult for anybody who is not a scientist of the very highest order to ever work out. . . ."

"That is right," Johnston responded.

Johnston was also asked if the eight hundred Boeing SSTs projected during the 1971 debate would have been a serious problem. He responded that they would have been "a worldwide disaster." He then added: "The fact that you had to ask that question is why we are here today. You had to ask me if the DOT really admitted that the economically justified number of Boeing SSTs would have been an environmental hazard. You should have known

that already by now. It is in here . . . is contained in this report but
not in those words. You have to read the fine print and multiply it
out. I am saying that Congress should understand that this report
reaches that conclusion."

In his testimony Donahue said the executive summary was "a
travesty, not addressing frontally the main issues CIAP was directed
to study and underplaying very serious difficulties that would
confront attempts to fly large numbers of SSTs. . . . The summary
was upbeat and positive in the traditional style of a report that is
designed not to damage or kill a program."

Donahue was particularly irritated by a table in the summary
dealing with the amount of ozone reduction that could be expected
from one hundred advanced SSTs. The figure, he said, was arrived
at by making several assumptions favorable to the SST that served
to reduce the calculated impact on ozone.

Donahue concluded that he would "dearly like to know" why the
CIAP findings were so well hidden and distorted. He referred to
news stories that indicated the State Department was trying to see
that the SST be given a fair shake. "Doesn't the State Department
know that the plane has already been given a fair shake and found
to be a dangerous device?" He pointed out that extreme caution
was being exercised in forbidding the use of fluorocarbons because
"we are dealing with a $5 billion industry already in existence,
manufacturing and using these propellants. Isn't it madness to
allow another such industry to be created in the face of its recognized threat to our environment? . . ."

Questioned by subcommittee chairman Randall, Steve Wofsy of
Harvard said that the technical details of the CIAP study had "not
been juggled or anything like that, but they are not usable for the
average person or even for a well-informed layman. . . . I found
it very difficult to get what I wanted out of those graphs even
though I have almost all of the numbers right in my head."

Robert Cannon was next to the witness stand. He asserted that
the CIAP report of findings dealt with the SST issue "with care,
skill, balance, perspective and total candor. It is a straightforward
report, in which all that is known is readily found and the remaining uncertainties are described meticulously." He said the skin-cancer statistics were dealt with late in the summary because it would
have been "grossly misleading and irresponsible" to quote the statis-

tics without a full and careful explanation of exactly what they meant. Finally, Cannon emphatically denied that senior Transportation Department officials had interfered with either the conduct of the study or the report of findings. The study was done "without any trace of coercion," he said.

The real fireworks at the hearing were provided by the cross-examination of Cannon by Randall and by Congresswoman Bella Abzug, then a member of the subcommittee. Abzug, a flamboyant feminist known for her outspoken manner and the wearing of large, floppy hats, was described by one CIAP insider who attended the hearing as a "tough Brooklyn scrapper." She asked Cannon why, since the CIAP study acknowledges that large-scale SST operations may occur in the future, the summary emphasized the insignificant number of thirty aircraft.

"Which you say first is a matter of which is better understood," Cannon said. "We decided, rather arbitrarily, to get the few-aircraft question out of the way first. One of the questions we were asked was, 'Are there any problems with the current fleet?' We answered that. Another question we were asked had to do with future fleets, and we answered that."

ABZUG: "I am a lawyer and I am very accustomed to answering what you want to answer and not answering what you do not want to answer."

CANNON: "We answered everything we were asked."

ABZUG: "We have these facts. They are before us. You have not sufficiently refuted them here."

CANNON: "Yes I have."

ABZUG: "I am not going to press this question because the record is clear."

CANNON: "I think the record is clear too."

ABZUG: "The difference is that we are going to be final determinants of that, not you."

Abzug then took up the matter of the Grobecker press conference at which the executive summary was released, which she described as an attempt to downplay and distort the actual findings.

"We did not distort anything," Cannon said.

"Why is there a discrepancy between the actual study, the executive summary, and the press reports?" Abzug persisted.

"There are no discrepancies," Cannon retorted.

Cannon, who had been assistant secretary of Transportation responsible for the CIAP program[3] was strongly challenged by Randall on his apparent lack of knowledge about details of the program. Cannon was unable to remember publication dates of CIAP documents and floundered when asked to name some of the scientists who had had CIAP contracts, referring both questions to Grobecker.

RANDALL: "You surely know? You know how many worked on the project, don't you?"

Cannon said there were between five hundred and one thousand around the world, and finally listed Stanford Research Institute and several government departments.

RANDALL: "Anybody else besides Stanford, outside of government?"

CANNON: "Of course . . . I would really like to look in the 'Report of Findings.'"

RANDALL: "Doctor, you worked on this for how many years? You can remember it. How many years did you work on it?"

CANNON: "The entire time until—"

RANDALL: "How many years?"

CANNON: "About 3½."

RANDALL: "Surely you remember the names without having to look them up in the book, don't you?"

CANNON: "I am thinking."

RANDALL: "I will give you a minute or two to think."

In the end, Cannon named Marcel Nicolet, who had been mentioned previously during the hearing, and Hal Johnston, who had testified that morning.

[3] At the time of the hearing, Cannon had left the DOT for a position with the California Institute of Technology.

Bella Abzug challenged Cannon to explain why the words "ozone" and "skin cancer" were not mentioned in the summary's conclusions or recommendations: "Why is ozone depletion played down and hidden in the technical data if you are so concerned . . . about prevention?"

"I think no one had any misunderstanding at all—and there were a lot of misunderstandings—but nobody missed the point that it was the ozone problem which was the No. 1 problem we were talking about," Cannon said. He said it appeared in every press article and more than half the summary was devoted to the ozone issue. "There was no intention whatever of underplaying . . . that subject."

Randall then asked Cannon if anyone in the Transportation Department or the executive branch told him how to word the summary. Cannon emphatically denied this.

"You were, however, aware of the fact of the interest of the executive [branch] in this issue," Abzug demanded.

Cannon said he was aware that interest in the SST went as high as the White House, but this had never been communicated to him in any direct way.

ABZUG: "Well, then, how did you know it?"

CANNON: "I read the papers."

ABZUG: "But it was not in the papers. We have only recently discovered the specific interest of, for example, the President—"

CANNON: "I believe the Administration's interest in the supersonic transport was pretty well documented."

ABZUG: "I know, but a special interest has been indicated—has been revealed—of ex-President Nixon in this SST—"

Abzug's questioning was interrupted by Randall, who asked if Cannon had been called for a conference with James Beggs, a senior Transportation Department official.

CANNON: "Yes."

RANDALL: "You are saying to us that he never discussed what his position on the SST was?"

CANNON: "No, I won't say that he never discussed what his position was."

RANDALL: "Are you saying that he did discuss with you his position on the SST?"

CANNON: "I am sure everyone around me was talking about how he felt."

RANDALL: "You are not answering my question."

CANNON: "Yes, of course he did."

RANDALL: "We may have to put you under oath here in a minute."

CANNON: "I'm sorry. Of course he talked about his position."

RANDALL: "Sure. He said he was interested in the SST?"

CANNON: "Of course."

Abzug resumed her questioning about the interest of former President Nixon in the SST.

CANNON: "I can't believe that I didn't read it in the Washington *Post* and everywhere else."

ABZUG: "I don't think so—not at the time, because it has taken me quite a bit of time to get the letters that were written by President Nixon."[4]

CANNON: "I think the Administration's position on the SST was very clear."

ABZUG: "Is this gentleman under oath?"

Cannon denied any knowledge of the Nixon letters. Abzug concluded: "You cannot dispel this feeling . . . that there was some effort to play down the actual facts as to what the scientific impact of the SST would be. Your demeanor, your testimony, the inconsistencies, the executive summary as against the report, and so on. This is what disturbs us."

At the conclusion of his testimony, Cannon said he had tried very hard to correct the impression left by the "very unfortunate interpretation" of the executive summary in the press.

---

[4] The letters had not even then been released.

Cannon deeply resented his treatment by the committee. He felt they had not done their homework and knew only what "a few very biased people had told them. They were . . . shooting in the dark." He felt that he was being treated like a criminal on trial.

Alan Grobecker, who was not called to testify, recalls the hearing with considerable bitterness. He was angered by the way Cannon was treated. "It was done in a dirty way. They asked him questions he was completely unprepared to answer. I was there, and they ordered me not to tell him. He had to ad-lib. My reaction was that it was a real political game they were playing, to discredit him and the CIAP report and develop anti-Concorde feeling."

Grobecker had a prepared statement, but he never got to deliver it. In it, he notes that any twenty-seven-page summary of a study totaling seventy-two hundred pages is going to omit many details. But he steadfastly emphasized that the summary "is a truthful description of the salient points of our findings and it is wholly consistent with the full report. . . . It fully, accurately, and without distortion describes the central conclusions of the study."

Grobecker's deep resentment of the charges against him was reflected in the rather injured tone of the last brief section of the statement: "I am extremely proud of the report and . . . I take personal exception to any possible implication that this report or any part of it was modified for political purposes or in any way attempted to misrepresent the facts as we understood them. . . . Reasonable men may differ as to what is the most accurate possible brief summary of a long report. But that is quite different from suggesting that any failings the summary may have had were deliberate or intended distortions. Criticism which impugns the motives or integrity of the report's authors is wholly unfounded and unwarranted."

The ozone controversy abounds with conspiracy theories, and none was more persistent than the one that postulated overt political pressure on the CIAP executive summary. But this particular theory was never proved, nor was a conspiracy necessary to explain the executive summary. The summary was indeed misleading in a way that favored the Concorde and SSTs generally. It did not represent the consensus of the CIAP scientists. But the explanation for its failings lies in the manner in which it was written and in the personalities of the men who wrote it.

The writing of the summary and, indeed, of the "Report of Findings" itself, was described by one who was close to the action as "six months of disaster." Cannon and Grobecker were writing, rewriting, and passing drafts back and forth. There were a dozen or more different versions. Cannon and Grobecker were constantly on the phone, sometimes for hours at a time, and one observer remembers that Grobecker could often be seen wincing as he listened silently to what was being said at the other end of the line. There were those associated with the project who, according to one source, would repair to a nearby bar to "sit around drinking beer and laugh about the crazy things that were going on" in writing the summary.

Cannon says that he kept himself carefully informed of the scientific developments in the CIAP study, although other sources familiar with the study said Cannon rarely attended internal meetings to discuss CIAP's progress.

Cannon was also deeply involved with a committee of the National Academy of Sciences which was doing an independent review of the CIAP data; he believed strongly that such an outside review committee was necessary and was instrumental in having the committee set up. In September 1974, the committee held a meeting on the CIAP data and Cannon presented a version of the executive summary. This document is labeled: "Very rough draft. Numbers are dummys [sic]. Do not use in any way." He noted the name of each person to receive one of the numbered documents and retrieved the copies later.

Some of those who heard Cannon's presentation were startled. Somehow, it had apparently escaped him that the prime focus of the CIAP study was the ozone problem, for the introductory part of his summary made only vague and inconclusive references to CIAP's findings with respect to ozone depletion. Moreover, he made the rather startling statement early in his summary that "the most important single finding of the study is that liquid water droplets from [high-flying aircraft] contrails will not persist in the stratosphere." The CIAP study had not addressed itself to the contrail problem in any major way at all.

"I almost fell right off my chair," said one observer at the meeting who heard Cannon's presentation. "It was clear he didn't know what the hell was going on."

Cannon, however, says the statement was in there because he was still making the transition from a primary concern about the climate problem to that about the ozone depletion problem. He added that the draft summary "clearly served its purpose" because others pointed out that it "didn't show good perspective, so we changed it. That's why I floated that draft."

In light of the flap that would later erupt over the executive summary's principal conclusions, there was one paragraph in Cannon's version that was particularly intriguing. His first principal conclusion read: "The results of the CIAP investigations to date, in general, support some of the environmental concerns which were the original impetus for this study." (Although he does not say so, those concerns were, of course, related primarily to ozone depletion.)

This statement, or one very similar to it, appeared in several of the earliest drafts of the executive summary, but it was quietly dropped somewhere along the line.[5]

The second conclusion in Cannon's paper is similar, but not identical, to the one that finally became the first principal conclusion in the official version of the executive summary. It stated that operations of present-day SSTs "do not create a significant perturbation of the *environment*." (Emphasis added.) The final version, of course, talked not about the environment, but about climatic effects.

There were those who put a sinister interpretation on this change, viewing it as yet another ploy to evade the ozone question. But this may simply have been the result of Grobecker's penchant for defining terms to suit himself; for example, subsuming ozone under climatic effects. He could also be very stubborn about these definitions. "Alan Grobecker made up his mind about certain things and absolutely refused to hear any dissenting views," said one coworker. As a result, he had a number of pitched battles with staffers, some of whom came to be known as the "CIAP underground."

---

[5] Grobecker categorically dismissed any conspiracy theories about the disappearance of this statement. He said it appeared in material written before he and Cannon really got down to hammering out the summary; Grobecker seemed to regard it simply as a casualty of the rewriting process. "It didn't seem that it was important enough a thing to bring out."

Grobecker denied that there had been political pressure from the White House. One member of the CIAP underground confirmed that this was true, at least as far as he could tell. "I never knew of anyone to even give a damn about the whole thing, much less try to influence it." But there may have been what he refers to as "perceived pressure." Grobecker had a background in military research (he was with the Institute for Defense Analyses before going to CIAP) and was, according to one source, "the kind of guy who says, 'Aye, aye, sir.'" He would be responsive to what he thought his superiors wanted to hear. Grobecker (and Cannon) "really did want to present a fair assessment, but they wanted to present it in a way that wouldn't upset people. They didn't want to come out with a report that was going to be digested as having said that the DOT says the SST can't fly because it will wreck the ozone layer. On the other hand, they didn't want to come out with a report that said the SSTs were no problem, because that was a lie. So they strove for what appeared to be a reasonable middle ground."

If political pressure did not really account for the failings of the executive summary, what did? At least part of the answer may be found in certain aspects of Grobecker's character. Perhaps the most important is that he is fundamentally in favor of the SST concept. He believes that people will pay for speed; they did when they switched from propeller-driven aircraft to jets, and they will again when they switch from subsonic to supersonic jets. Grobecker envisions a supersonic age; he refers to Coleman's decision to allow Concorde flights to the United States as "statesmanlike" (because, he laughingly adds, it is the decision he would have made). Gazing at the small, sleek model of the Concorde on his desk, Grobecker remarked with some vehemence: "It would have been a crime to have killed off this airplane with the small numbers flying."

And then there was Grobecker's technological optimism, a description of himself he readily accepts. He regards the idea of banning SSTs as a variation of Ludditism. His technological optimism was shared by many of the senior people in the Department of Transportation and in the White House; it permeated their public statements and probably their thinking as well. They attacked the skin-cancer issue as an instance of "irrational fear in the face of

technological change." The view that harmful environmental consequences of the SST could be avoided or controlled by technology
prevailed. Cannon stressed the need for preventative measures to
ensure that there would be no harmful side effects, saying, "I believe our heritage is to focus on how to solve problems." Fred
Singer, head of the DOT's SST Advisory Committee, said he
believed that to overcome the adverse effects of technology "you
need more of the right kind of technology."

These are simply not the type of people who think seriously in
terms of the option of not flying SSTs; they think, instead, of what
can be done to ensure that the aircraft can fly. This emphasis on the
technological fix is hardly surprising coming from the Department
of Transportation, an agency committed to advancing the state of
aviation technology.

In its coverage of the Randall subcommittee hearings, *Science*
magazine noted that though the chairman had seemed to be fishing
for evidence of bad faith, "finding an explanation for the fact that
the ozone and skin cancer problem were addressed obliquely in the
summary does not require questioning anyone's motives. If Congress
expected a report couched strictly in terms of potential environmental problems—with little emphasis on how future SST technology might overcome them—it would have done better to put the
Environmental Protection Agency or the National Oceanic and
Atmospheric Administration in charge of CIAP."

The executive summary could thus have been the result of no
more than a desire to put a positive rather than a negative face on
things, a desire to give the SST the benefit of the doubt.

Of course, there were those who didn't believe the SST deserved
such a break. Mainly, their attitude toward the SST was: Who
needs it? This, of course, is a value judgment, just as much as were
the views of the pro-SST camp. Thus the disagreement over the
CIAP executive summary can be seen primarily as another in the
long line of increasingly fractious philosophical disputes between
those who are committed to technological "progress" and those
who would slow or stop these developments for fear of their adverse
environmental consequences.

These divergent views were clearly expressed in a series of letters
to *Aviation Week and Space Technology* on the Concorde issue. "I,

for one, intend to resist being stampeded by vested interests and their dupes, the progress worshipers," wrote one reader.

Another said: "Being an electronics engineer, I long held the view that advancing technology for technology's sake was in the better interests of mankind; in fact, I was highly disappointed when Congress voted down the domestic SST program. I have now come to realize that the benefits of new technology must be carefully weighed against its pitfalls. . . . There must come a time when mankind must decide where to stop unleashing new technology without an adequate understanding of its potential harm. The time is now and let us start with the Concorde. . . ."

On the other side was the reader who divided the Concorde dispute into two camps—"the passionate, emotional pseudoscientific nonsense perpetrated by almost all of those on the negative side of this overblown controversy and the rather cool, almost detached presentations offered by the pros."

After the November 13 session of the Randall subcommittee hearings, the CIAP study and the ozone question were removed from center stage. On November 14 and again in mid-December of 1975, the subcommittee was preoccupied with an investigation of possible political pressure on the Environmental Protection Agency and the Federal Aviation Administration and with persistent but unsuccessful attempts by Bella Abzug to wrest copies of the Nixon letters from officials of the EPA and the FAA. While she had him as a witness, Abzug asked EPA Administrator Russell Train about an ad that the Concorde manufacturers had taken in the New York Times. "The ad says a thousand scientists worldwide concluded that the proposed Concorde services [to the United States] would have no measurable effect on the stratosphere and they emphasize the word 'no.' Do you consider that a fair and accurate statement in light of what we have heard since about ozone depletion and skin cancer?"

"It certainly is not a good statement," Train responded. ". . . There are substantial unresolved problems involving upper-atmosphere effects and they certainly should be resolved before any commercial fleet operations of the SST should ever be approved. . . . No increase in the incidence of an adverse health effect is so negligible that it should not be a matter of some concern. . . . A determination has to be made as to whether that incidence of risk is really so significant that the aircraft should be banned. . . ."

ABZUG: "I do not believe that human life is expendable for an aircraft."
TRAIN: "We expend fifty thousand people a year for an automobile."

Finally, during testimony on December 12, Transportation Secretary William Coleman released copies of the Nixon letters. He said he had not seen them before that morning and vowed that his upcoming decision on Concorde landing rights would be based entirely on the public record and not on secret diplomatic or political commitments.

The letters assured the French and British leaders that the Concorde would be treated fairly so that it could compete for sales in the United States "on its merits." It also promised that a noise regulation then being proposed by the Federal Aviation Administration would be issued "in a form which will make it inapplicable to the Concorde."

SST opponents of course immediately seized on this with an I-told-you-so attitude. As it turned out, the FAA decided not to institute the noise rule anyway at that time, for quite different reasons (they thought it unworkable). At the Randall hearings, Coleman was asked if he thought Nixon's "commitment" on the matter was therefore without any real meaning, and Coleman agreed.

Concorde opponents were then reduced to arguing that the letter went to great lengths to convey the *impression* that Concorde would receive favorable treatment. "I don't read it that way," Coleman responded blandly; to him, it was a straightforward and perfectly acceptable assurance of fair treatment and no more.

In fact, the letter was artfully worded to appease the British and French while at the same time carefully avoiding commitments that could not be made. Coleman aptly but unkindly characterized it as being "full of sound and fury, signifying nothing."

At the end of December, the World Meteorological Organization (WMO) issued a statement on the modification of the ozone layer due to human activities. The section on SSTs was curiously reminiscent of the CIAP executive summary. The first conclusion was that currently planned SSTs—a projected fleet of thirty to fifty aircraft flying at an altitude of about seventeen kilometers—would not have a significant or even distinguishable effect on the ozone layer. The second conclusion was that a large fleet flying at greater altitudes would have a noticeable effect and that international agreements might be necessary to limit total emissions of pollutants into the stratosphere. A WMO spokesman in Geneva stressed that the statement was not intended as an endorsement of the Concorde, which is not mentioned by name in the report, although the projected fleets were based on the Concorde and the TU-144. But the timing of the release of the statement and its emphasis on the relatively minor ozone threat to be expected from the current SST fleet could hardly have been more fortuitous for the Concorde, which was then undergoing a series of proving flights. Some press accounts interpreted the WMO statement essentially as a green light for the Concorde.

As 1975 drew to a close, the fight over U.S. landing rights reached a fevered pitch. In November, when he had released the final environmental-impact statement on the Concorde,[6] Coleman had announced he would hold a one-day hearing before he rendered his decision. Coleman, a lawyer and former Board member of Pan Am, had taken over as Transportation Secretary only the previous March, and the politically sensitive Concorde decision had been, as the Washington *Post* put it, "bucked upstairs" to him.

The imminence of Coleman's decision provoked French and British unions into overt and rather shrill threats of retaliation if Concorde were refused permission to land in the United States (for example, they threatened to refuse to service U.S. planes at European airports). Official reactions by the French and British governments were somewhat more circumspect, but there was a good deal of open speculation about the retaliatory means at their disposal.

The standard British attitude that the Americans *will* get too worked up about the environment also surfaced. There seemed to be a resentment against what the British magazine *New Scientist* re-

[6] The final statement, avoiding the pitfalls of the CIAP executive summary, used both the National Academy's SST study and the full CIAP "Report of Findings" as its references on the ozone problem.

ferred to as "environmental neocolonialism." The editorial, written by Jon Tinker, suggested that, because of its economic clout, the United States is able to "export its environmental conscience."

At the same time, opposition to the Concorde from its political opponents and environmentalists was equally emotional, strident, and intractable. In view of the mounting pressures, it is little wonder that Coleman opened his one-day hearing on January 5 by saying: "Of course, I do not expect to decide this issue in a way that will please everyone. I only hope that by having the issues aired in this forum, all will recognize that this decision will be made without prejudgment or bias, absent of any prior commitment to any person, organization, or government."

Coleman heard about sixty witnesses, a scrupulously equal complement of Concorde opponents and supporters. The former predictably raised the ozone issue; the latter just as predictably condemned it, and clearly no one's mind was changed. The *New Scientist*'s Peter Gwynne, in fact, said that the hearing "resembled nothing so much as the latest in an interminable series of dull cup replays between two evenly matched football teams—neither of which has the spark of brillance necessary to gain a decisive victory." Thus, while the hearing may have appeased Coleman's evident desire for a public demonstration of his impartiality and open-mindness, it did little by way of producing substantive new information on the ozone question. With the decision due early in February, the comments from both the pro-SST and anti-SST factions became, if anything, even more shrill than before. *Aviation Week and Space Technology* ran a strident editorial whose basic premise was that the Concorde deserved its chance. It was written by publisher-editor Robert Hotz and was reminiscent of the 1971 editorial he had written lambasting Jim McDonald for his skin-cancer theory. "Never in the history of aviation has such a mass of political vituperation, mass hysteria, and technical baloney engulfed a new aircraft as that which has surrounded the . . . Concorde. . . . The 'Concorde causes cancer' war cry has been the most hysterical and flagrantly false premise of the anti-Concorde mob."

On January 21, 1976, just two weeks before Coleman was to announce his decision, British Airways and Air France lavishly and flamboyantly inaugurated commercial Concorde service, officially opening the era of supersonic passenger travel. The two planes took off simultaneously at 12:40 P.M. GMT from London's Heathrow Airport and Paris' Charles de Gaulle Airport. To preclude any possibility of one-upmanship, the two pilots went through a thirty-second countdown on their respective runways. The Air France jet departed for Rio de Janeiro, while the British Airways plane headed to the Middle East country of Bahrain. On board the British aircraft were Sir George Edwards, retired chairman of the British Aircraft Corporation (builders of the Concorde), who had left a sickbed to make the trip, and Lord Leathers, a director of the National Westminster Bank, who had reserved the first Concorde ticket twelve years before.

Not surprisingly, the ballyhoo attracted a good deal of press attention, and readers were regaled with such minutiae as the fact that, on the Rio to Paris run, one passenger's pet became the first supersonic dog. The fare for the dog was $245.[7]

Coleman's decision, released on February 4, paid considerable attention to the ozone problem, even though he had concluded that the impact of the flight schedule he was considering (six daily flights for sixteen months) would be negligible.

The Secretary seemed less than totally persuaded of the link between ozone depletion and skin cancer, for he emphasized the uncertainties and the fact that there was "genuine dissent at each point in the causal chain." Despite the lack of conclusive proof, he nevertheless accepted the theory as correct and therefore could not ignore the possibility that the Concorde flights might cause some increase in skin cancer. Coleman also recognized a point that is too often ignored by SST supporters: SST operations could subject *everybody,* whether they choose to fly in SSTs or not, to an increased risk of skin cancer. SST proponents were fond of pointing out that the risk would be no greater than that caused by a little extra sunbathing or by moving South, but they failed to recognize that these latter activities, and the attendant risks, are undertaken voluntarily

[7] This story elicited a letter to *Aviation Week* from one E. D. Francis, who claimed the first supersonic rattlesnakes (carried on an F-4B flight), and another from England telling of a supersonic sausage carried on a Concorde test flight.

by individuals and for themselves only. Moreover, people can choose not to subject themselves to these risks, something they cannot do in the case of SST operations.

In the end, however, Coleman concluded that even though there might be a slight risk of additional skin cancer, the risk was not sufficiently large to deny the Concorde its sixteen-month demonstration period, "the stratospheric impact of which would be miniscule."

He rightly pointed out that the United States harbored several other potential threats to the ozone layer—fluorocarbons, the space shuttle, fertilizers, military SSTs—and had not, up to that point, taken any action against them. "Unless the United States is willing to act against these sources as well, action against the Concorde could well be perceived as discriminatory. Such discrepant treatment would be justifiable only to ward off a substantial and immediate danger of harm and the danger posed by these flights does not, in my opinion, fall into this category."

Finally, Coleman argued that the United States would be affected no matter where the Concorde flew in the Northern Hemisphere, and by participating in the program, the country would be in a better position to press for international standards and improved SST technology to protect the world from significant harm.

"I have enough confidence in this nation's environmental commitment and in the objective judgment of the marketplace to be sure that if the SST does in fact become the aircraft of the future, it will only be because man will have developed the technology to meet rational environmental standards and to enable the SST to compete in the marketplace effectively. But if we bar Concorde completely, we may well be condemning for all time or delaying for decades what might be a very significant technological advance for mankind."

Coleman concluded that "the benefits of an environmentally sound, commercially viable SST would be substantial" and that the U.S. demonstration test "is needed to determine whether a commitment to this new technology should be embraced."

The decision came as a shock to many people; the expectation had been that Coleman would allow flights into Dulles only for six months to a year. Coleman had final say at Dulles because that airport is operated by the Department of Transportation, but it was not at all clear that the Secretary's will would prevail at Kennedy Airport, which was operated by the Port Authority of New York and New Jersey. Landing rights in New York were particularly important to the Concorde, and there was some speculation that Coleman decided to throw this hot potato into the states' lap, since the federal government had to live with Paris and London, while the Port Authority did not.

The New York *Times* and the Washington *Post*, never enthusiastic Concorde supporters, were, however, cautiously approving of Coleman's decision. But they emphasized the "trial period" theme, the *Times* call it an "explicitly tentative ruling." In his own inimitable style, CBS-TV pundit Eric Sevareid was ambiguously complimentary. Calling the Coleman decision a "case of muddling," he added that it "took courage to muddle." Coleman, he said, had the "courage of his doubts" and had made a "decision not to decide."

From the SST's foes, however, there was no ambiguity: Environmental groups, congressional opponents, New York politicos, and citizens' groups were all predictably outraged. "The decision is not just a tragedy, it's a fraud," said Congressman Herbert Harris of Virginia.

Congressman Lester Wolff of New York said that the decision made a mockery of the Environmental Protection Agency and vowed that Coleman would be called before House committees to explain himself.

On February 24, about three weeks after he released his Concorde decision, Coleman again appeared before the Randall subcommittee, where he got into a spirited and rather heated debate with Bella Abzug. But Abzug was not able to browbeat Coleman, as she had some of the other witnesses; Coleman, who had done his homework, held his own against the formidable New York congresswoman.

In response to a question, Coleman said that the CIAP and National Academy of Sciences reports on the SST indicated that, in dealing with the small number of SSTs covered by his decision, "there is no measurable effect or change in the ozone. It is also my understanding that everything you say is a theory. . . . You cannot keep on talking about these effects when you realize that there is countervailing evidence and that the theory has not been borne out entirely in practice."

"I rest my case on science," Abzug replied, "and I do not think that Americans should be used as guinea pigs. We must first develop a clean SST before we unleash it upon the people in the country or in the world."

But Coleman had the last word. "I think you have to be impressed by the fact that the world scientists, when they looked at this problem, said that up to sixteen SSTs would produce no measurable effect whatsoever. . . . I do not see how you, in good conscience, could say that you would bar something which everybody concedes, even those who hold the theory as strongly as you do, that this number will not have any effect."

Concorde opponents launched a series of legal and legislative attacks on the Coleman decision but failed to overturn it. However, New York State and the Port Authorty of New York and New Jersey imposed a ban for Kennedy Airport and made it stick for nearly two years. Concorde service to Washington officially began on May 24. Just two weeks later, the French and British announced that they were ceasing production of the Concorde after the first sixteen were built; they could not sell even all of these first-run aircraft. Increasingly convinced that New York service held the key to economic success, they continued to fight the ban there. It was estimated that Air France spent as much as $2.7 million to $4 million on its legal, public relations, and lobbying effort, while the British, more modestly, spent less than $200,000.

For nearly two years, the Concorde battle raged on, wending its inexorable way through federal and appeals courts—and a bewildering array of yeas and nays—until finally, in October 1977, the Supreme Court voted yea. In November the French and British began commercial flights into Kennedy Airport, their make-or-break market.[8]

A large part of the debate over the Concorde related to the prob-

---

8 Earlier, in September, the Carter administration had proposed that the existing Concordes be allowed to land in at least thirteen U.S. cities unless prohibited by local authorities on the basis of "reasonable, nondiscriminatory noise rules." (In other words noise rules that applied equally to all aircraft, not just the Concorde.) In addition to New York and Washington, the cities included Philadelphia, Anchorage, Boston, Dallas, Honolulu, Los Angeles, Miami, Houston, Chicago, Seattle, and San Francisco.

lem of discrimination. The United States, which supplies 90 to 95 per cent of the jet aircraft flown in the non-Communist world, has been the prime beneficiary of an international aviation system founded on reciprocal and nondiscriminatory rights and privileges. The Europeans resentfully believed that a refusal to allow the Concorde to land in the United States was an out-and-out denial of these rights; never very excited about the ozone problem, they believed that the environmental arguments were simply excuses—a smoke screen behind which the United States could hide its real motive, which was to protect its own airlines and aviation industry.

In truth, the ozone issue was a difficult one to use in the fight over Concorde landing rights for a few planes for sixteen months. And Coleman had made it clear that if the Europeans wanted to schedule more flights or establish permanent operations, a further environmental-impact statement would be required that would take into account the effects of such expanded operations.

Many SST supporters argued that it was perfectly reasonable to give the Concorde a "fair test." A French official said: "This period will allow the plane to confirm what we already know—that its impact on the environment is negligible." Apart from the fact that this was prejudging the results of the test, this argument was patently false insofar as the ozone problem was concerned. The trial period would not constitute a fair test of the Concorde's impact on ozone; the number of planes was too small, and the time was too short for that.

On the other hand, Concorde foes were worried that an SST service on the busy and lucrative North American run, even for a short period of time, could generate considerable consumer demand for supersonic travel, and this, in turn, might open the door to a large fleet of advanced SSTs. "We're just kidding ourselves if we think this whole debate is over sixteen months of SST flights into Dulles," said Republican Senator J. Glenn Beall of Maryland. The sentiment was summed up by a Herblock cartoon showing a malevolently grinning Concorde "mother bird" hatching a brood of baby SSTs.

The prevalent view was that these second-generation SSTs would more than likely be born in the United States. As Allen Hammond wrote in *Science* magazine, there is "an assumption widely held in the U.S. aviation industry—that, notwithstanding congressional decisions, and gloomy economic prospects to the contrary, there will eventually be a U.S. supersonic fleet."

There is no government commitment to build an SST, but the U.S. aerospace industry is keeping its hand in, with the aid of federal government funds. The National Aeronautics and Space Administration has, since 1972, funded SST studies by several aerospace companies and engine manufacturers under the Advanced Supersonic Transport (AST) program. One of the companies, Douglas Aircraft Company, has also provided its own funds to do a preliminary design of an advanced SST.

At $8 million to $9 million a year, NASA supports design studies —mostly an exercise in keeping an eye on technological developments. This funding is certainly not enough to move the program from paper to hardware; nevertheless, the aerospace industry seems keen to do this and, at least publicly, expresses considerable confidence that it will eventually happen. In fact, according to Richard FitzSimmons, director of Douglas's advanced SST studies, American aviation may have "a new Sputnik to face" if U.S. Concorde service proves successful. Raising the specter that propelled the United States on its headlong rush to the moon so as to trample the Soviets in the dust, FitzSimmons notes: "Our airlines, our industry and our nation will have been put again into a position of having to catch up."

It remains uncertain just how much of a threat to the ozone layer the SST will ultimately be. Part of this uncertainty stems from economics, which will determine just how many planes will fly. The Concorde production lines have been shut down, and it is uncertain whether a cleaner, quieter advanced version of the needle-nosed craft will emerge. But the main factor affecting the projected impact of SSTs on the ozone layer has to do with the continuing refinements in our knowledge of atmospheric chemistry. In mid-1977, more than two years after the completion of the CIAP study, a new set of computer calculations indicated that SSTs, far from destroying ozone, would probably have a negligible effect on the ozone layer.

How did this happen? As we have seen, the earth's atmosphere is chemically a very complex place; the turnaround in the SST story resulted from the fact that injecting nitrogen oxides into the atmosphere can *either* create or destroy ozone (something scientists had known prior to the CIAP study). Near the ground, in urban air containing hydrocarbons—such as exists in many large cities— NO_x emissions from automobiles result in the *production* of ozone, one of the most noxious ingredients of smog.

As you go higher in the atmosphere, however, NO_x switches from an ozone-producer to an ozone-destroyer. As we have seen, the CIAP scientists believed that, in the stratosphere, the major effect of NO_x would be to destroy ozone; in fact, they believed that the altitude at which the switchover occurred was around ten kilometers—well below the levels at which the SSTs would be flying.

However, two things have since happened to change this picture. First, the computer models now include chlorine in the stratosphere from both natural and man-made sources (whereas, the models used in the CIAP program did not account for chlorine, whose existence in the stratosphere was unknown at the time).

Second, improvements have been made in laboratory measurements of the speeds of chemical reactions used in the computer models of the atmosphere. Of particular significance was the finding by Carl Howard and his coworkers at the Aeronomy Laboratory of the National Oceanic and Atmospheric Administration that one of the key reactions was actually some forty times faster than had been previously believed.

Both of these factors had the effect of increasing the altitude at which the ozone-producing effect dominates over the ozone-destroying effect. The "switchover" point is now thought to be at about the same level as the SST's cruising altitude, and thus it appears that the aircraft will have little effect one way or the other on the ozone layer.

The SST problem was not the only one to be significantly affected by the new measurements and calculations. As we will discuss in more detail later, the new data also resulted in substantial changes in predictions of ozone depletion to be expected from fluorocarbons. However, in this case things went in the opposite direction; that is, it appears that fluorocarbons are twice the problem originally calculated.[9]

[9] For a discussion of the political dilemma this creates for scientists, see the Epilogue.

If the first phase of the ozone story involved the SST, the second involved another and even more exotic flying machine—the space shuttle. The shuttle is an airplane-like vehicle about the size of a DC-9. It is designed to be launched vertically like a rocket, to operate in orbit like a spacecraft, and to land horizontally on a runway like an aircraft. The first production model was rolled out in late 1976 and has been undergoing tests in preparation for its first orbital flight in 1979 or 1980. (This vehicle was originally named *Constitution*, but it was changed to *Enterprise* after a letter-writing campaign by thousands of "Star Trek" fans demanded that it carry the name of Captain Kirk's starship in the old TV series.)

The shuttle, surprisingly enough, is the link between the SST and the fluorocarbon controversies. Unlike the SST, it does not produce NO_x—it produces chlorine. It was, in fact, the first man-made source of chlorine to be studied, and this paved the way for much of the fluorocarbon work that followed.

The shuttle debate started at a meeting held in Kyoto, Japan, in 1973—a meeting that very nearly didn't happen because of the ongoing territorial war over the regions of the earth's atmosphere.

CHAPTER SIX

Shuttle Diplomacy

Once upon a time there lived in the land of *IAGA,* in the kingdom of *Aeronomy,* strange creatures called *aeronomers.* Little was known about these creatures because they lived most of their lives in the remote areas of the kingdom, more than 60 kiloleagues from Earth.

Not so long ago a part of their kingdom, known as *Stratos,* was threatened by the invasion of a flock of big birds who make noises, that sound something like—sst. Some of the creatures of *Aeronomy* rushed to *Stratos* to try to discover what these birds might be doing to their kingdom. Some came because they heard these birds could also lay golden eggs.

We soon learned that there are three kinds of *aeronomers.* There is a group of high priests called *modelers.* They never go outside their temples where they try to prophesy what the big birds will do by examining the entrails of large animals called *computers.* Another group, who appear to be the worker drones called *experimenters,* spend most of their time in noisy, smelly rooms called *laboratories* playing with little boxes whose purpose seems to be the generation of random numbers called *data.* A strange relationship exists between the *modeler* priests and the *experimenters.* The priests feed the *data* to the *computer* animals and then study their entrails. They then tell the *experimenters* what kind of new *data* the animals need and the *experimenters* rush back to their *laboratories* and make more black boxes.

The third group is called *observers.* They also make black boxes but they throw their boxes into the sky. Most of the time the black boxes break. Sometimes they too give *data* which the high priests also give to their animals. However, the animals sometimes get sick if they eat this *data* and may even die if too many different kinds of *data* are fed to them at the same time. However, the high priests have become very clever at getting their animals to accept almost any-thing.

The diet of these animals seem to lack one essential nutrient called *transport data*. Unfortunately, these *data* are grown mostly by *dynamicists* who live in the land of *Tropos*. Only recently have the borders between *Tropos* and *Stratos* been opened to allow *dynamicists* and *aeronomers* to talk to each other.

—HAROLD SCHIFF, opening remarks,
IAGA meeting, Kyoto, Japan,
September 1973.

The problem of chlorine in the earth's atmosphere was first seriously discussed by experts in the fields of stratospheric chemistry and physics in September 1973. The scene was an international scientific meeting in Kyoto, Japan—a meeting that very nearly didn't happen, for reasons referred to only obliquely in the last sentence of Schiff's opening remarks.[1]

The Kyoto meeting had, in fact, precipitated an outbreak of hostilities within the international scientific associations. It was the latest round in the continuing jurisdictional dispute over the regions of the earth's atmosphere. Blockades were figuratively thrown up around the stratosphere. There was talk of invasion. There was talk of collision. There were summit meetings to resolve the issue.

It all started when Marcel Nicolet suggested that the meeting in Kyoto include a symposium on the aeronomy of the stratosphere and the mesosphere (the region above the stratosphere). Nicolet, a Belgian, was one of the founders of aeronomy (the science of the atmosphere) and a power in international scientific affairs. He made his recommendation to the sponsors of the Kyoto meeting, the In-

[1] Schiff was editor of a special volume of the *Canadian Journal of Chemistry* in which the proceedings of the Kyoto meeting were to be published. He was urged by several people who heard his remarks to include them in the proceedings, but this did not meet with the approval of the editors of the Canadian journal, who thought the remarks too frivolous for a serious scientific document. In consequence, they were removed from the regular editions of the Canadian journal, but Schiff insisted on their inclusion in the copies that *CJC* was publishing under contract to the CIAP program.

ternational Association of Geomagnetism and Aeronomy, known as IAGA. IAGA is one of the associations of the International Union of Geodesy and Geophysics, known as IUGG. The executives of both groups agreed to the symposium, and Nicolet proceeded to organize it.

His activities immediately provoked the wrath of another group, the International Association of Meteorology and Atmospheric Physics, known as IAMAP. This association, composed primarily of meteorologists, was already planning to hold a meeting on the stratosphere in Melbourne, Australia, in January 1974. IAMAP officials were incensed that IAGA was contemplating an earlier meeting on the same subject; they considered this to be in the nature of a preemptive strike. IAMAP had staked out the troposphere and the stratosphere, relegating IAGA to the mesosphere. They insisted that IAGA not cross the territorial borders. The fences were up. Passports were required.

A series of angry letters began flashing back and forth. One IAMAP official protested IAGA's activities. IAGA was proposing to "invade those portions of the atmosphere in which IAMAP has long been recognized as the competent body." He warned that a serious situation threatened to erupt.

A suggestion by Nicolet that IAMAP cosponsor the Kyoto meeting was rejected by IAMAP officials. Affronted, Nicolet pre-emptorily quit the Kyoto meeting. The excitable Belgian had been so incensed by a letter he had received from another IAMAP official, Will Kellogg of the National Center for Atmospheric Research in Boulder, Colorado, that he also announced that he was withdrawing from a commitment to give the first Sidney Chapman lecture at the University of Colorado in Boulder. Sidney Chapman, as we have seen in Chapter Two, was the father of aeronomy, and being offered the Chapman lectureship was a significant honor; it proclaimed Nicolet the inheritor of Chapman's mantle. But Nicolet would not set foot in Boulder as long as Kellogg was there.

At this point, Schiff reluctantly inherited the chairmanship of the symposium and the political mess it had wrought. Another Belgian scientist, Marcel Ackerman, was deputized by a group of Nicolet's friends and colleagues to visit Schiff in Toronto to urge him to accept the chairmanship and, in particular, to find a way of inveigling

Nicolet back into the fold. This was accomplished by inviting Nicolet to give the keynote paper at Kyoto.

Schiff had put a condition on his acceptance of the chairmanship of the symposium: A way had to be found to defray the large travel costs the invited speakers would incur in getting to Japan. It was the only way to ensure that the best scientific talent could attend. Schiff spoke to Alan Grobecker to see if CIAP funds might be made available for this purpose.

Grobecker was enthusiastic. He thought the timing was right. The CIAP program had reached the halfway mark, a lot of data had been collected, and a major meeting on the stratospheric work seemed appropriate. Grobecker was equally enthusiastic about the IAMAP meeting scheduled for Australia in January, and he saw no conflict between the two. IAGA could concentrate on the chemistry of the stratosphere and the refinement of the computer models used to predict stratospheric effects; IAMAP could concentrate on transport—air motions—and on the meteorological aspects of the problem. To keep peace, he gave the squabbling children $50,000 to help with travel. And so, on September 10, the scientists gathered in Kyoto.

The issue of chlorine chemistry in the earth's stratosphere was introduced at the Kyoto meeting by Richard Stolarski of the University of Michigan. Although he did not talk about the shuttle (for reasons we'll discuss later), this work had been done by a team of researchers at Michigan under contract to the National Aeronautics and Space Administration (NASA), which had become concerned about the possible stratospheric impact of the shuttle.

The story of the space shuttle really started back in early 1972, when NASA announced that solid-propellant rocket motors had been chosen to boost the shuttle into orbit. At that time, no one worried too much that the propellants would produce hydrogen chloride (HCl), which could be broken down in the stratosphere to produce chlorine atoms.

In July 1972, NASA's final environmental-impact statement on the shuttle revealed that HCl would spread along the shuttle's trajectory from ground level to the upper atmosphere, with the largest amount being deposited directly into the stratosphere. The projected fifty flights a year would dump some fifty-five hundred tons of HCl into the stratosphere annually.

The environmental-impact statement devoted considerable attention to the effects of HCl in the regions of the atmosphere below and above the stratosphere, but not in the stratosphere itself; there was no indication that the stratospheric effects of HCl were even considered and rejected as unimportant. Interestingly enough, two other shuttle effluents, carbon dioxide and water vapor, *were* considered. The impact statement concluded that neither would have unacceptable stratospheric effects and that "no negative environmental effects in the stratosphere are expected as a result of shuttle operations." It ended, on a grand note, claiming that shuttle-launched satellites will help to improve the management of the environment and the earth's natural resources. "This nation's short-term investment in the space shuttle program will result in a long-term improvement of the global environment for future generations."

The environmental-impact statement's omission of stratospheric HCl effects was seemingly so blatant that one is tempted to conclude that it had to be deliberate. But a more likely explanation is that NASA simply did not have much expertise on the subject of stratospheric chemistry at the time. Ron Greenwood, later head of NASA's stratospheric research program, adds that the HCl problem was ignored because "it was unheard of. In hindsight, it was obvious that it should have been picked up, but the impact statement was prepared a year before anybody was seriously thinking about the possible effects of chlorine."

In fact, the first draft of the impact statement was released before even the SST threat to the ozone layer had been widely recognized. When the final statement came out in early 1972, stratospheric chemistry was still not well understood, and certainly the problem of chlorine in the stratosphere had not become an issue in scientific circles. Though many scientists outside NASA knew that chlorine atoms were ozone eaters, few if any knew that the shuttle would be a source of stratospheric chlorine.

However, the impact statement left many questions unanswered, and a number of NASA scientists were dissatisfied with it. Even before it was officially released, a reinvestigation of the shuttle's environmental effects was under way. NASA's Marshall Space Flight Center awarded a contract to a team of researchers at the University of Michigan to consider environmental effects that may have been missed. At about the same time, a Shuttle Exhaust Effects Panel was set up within NASA with Ron Greenwood as chairman.

Near the end of 1972, Rich Stolarski, who is now with NASA but was then a member of the University of Michigan team, went to spend a year's sabbatical with Bob Hudson, a NASA scientist at the Johnson Space Center in Houston. Hudson, Stolarski, and Ralph Cicerone, another member of the team back in Michigan, discussed the shuttle/chlorine work on the phone about once a week.

Cicerone and Stolarski were yet another team of "outsiders" when it came to problems of the ozone layer. Like Rowland and Molina, they were not stratospheric chemists. Unlike Rowland and Molina, they were not even chemists. Cicerone had received his degree in electrical engineering and both he and Stolarski, whose training was in physics, had been doing work on the ionosphere, a region of the atmosphere above the stratosphere.

Stolarski remembers that they were looking around for something new to do and were attracted by all the activity in the stratosphere. But it was not easy to break into that game. They were not in the CIAP program and "you couldn't just hop into the SST thing. It had been going for too long and we didn't have the credentials." So they decided to take "something that nobody cared about, which was chlorine. It looked like a nice, *quiet* piece of the stratosphere to cut off and maybe get a paper or two while we were learning and nobody would bother us." In fact, they saw the shuttle contract as little more than a means of obtaining the money to begin this research.

It was through conversations with Don Stedman at Michigan that Cicerone and Stolarski first got the idea that ozone might be affected by chemical reactions involving HCl and its decomposition products. Stedman, a British chemist, advised Cicerone to write to Stedman's former research director, Michael Clyne, for information about stratospheric chlorine chemistry. Clyne wrote back with the rate constant (speed) of the reaction between atomic chlorine and ozone. This was the first half of the chlorine chain, but the Michigan group did not immediately realize that in fact they were dealing with a chain. It was near the end of 1972, during one of the three-way conversations among Cicerone in Michigan and Stolarski and Hudson in Houston, that they caught onto this fact. Initially, however, they got the wrong chain. The first part—the reaction between chlorine atoms and ozone—was right, but they closed the chain with another reaction that turned out not to be important.

Unknown to the Michigan group, the shuttle/chlorine problem had already been worked out by a team of researchers at the Lockheed Palo Alto Research Laboratory. The team, headed by Hiro Hoshizaki, had done this work on a contract from the CIAP program. They had been asked to study engine exhausts, and CIAP manager Alan Grobecker told them to include in their study the exhausts from rocket engines, including the shuttle. They got onto the chlorine question because Hal Johnston had alerted them to the potential of the chlorine chain for ozone destruction.

According to one member of the Lockheed research team, they tried to interest NASA in their work in September 1973; they were in Washington talking to the CIAP people and took the opportunity to try to set up a meeting with NASA. The attempt failed; their top management reportedly phoned them in Washington and told them to back off. It was an awkward situation; the Lockheed Propulsion Company was, at that time, competing for the contract to build the shuttle booster engine. In November 1973, the contract went to Thiokol Chemical Corporation of Utah (the home state of both James Fletcher and Frank Moss, at that time NASA administrator and chairman of the Senate Space Committee, respectively). In January 1974, Lockheed challenged the award of the $106 million contract to Thiokol on economic grounds. The challenge was not successful and it was only after the dust settled were the Lockheed scientists permitted to discuss their results formally with NASA. By March 1974, their funding from CIAP had run out, and they tried to get NASA support to continue their research. NASA was not interested; Hoshizaki got the impression that the space agency preferred to have the work done by university researchers on the grounds that research by aerospace industry scientists might be considered somewhat suspect in congressional hearings. So the team, which did not really have a good computer modeling capability anyway, published their results in their CIAP report and dropped the whole matter.

By that time, the chlorine issue was already being discussed within NASA by three space agency scientists: Bob Hudson at the Johnson Space Center, Jim King at NASA's Jet Propulsion Laboratory, and I. G. Poppoff of NASA's Ames Research Center. And in the spring of 1973, the Michigan group made its first formal presentation to NASA's Shuttle Exhaust Effects Panel and later submitted a written report (still containing the incorrect chlorine chain) that concluded that, on a global scale, chlorine compounds "may be significant destroyers of ozone. . . ."

Through the spring and summer of 1973, NASA started putting together a program to study the shuttle's atmospheric effects, and senior officials in the office of manned spaceflight and at NASA headquarters were informed of the problem. By mid-1973, according to Cicerone, NASA began leaning on the Michigan scientists to keep quiet about the shuttle chlorine problem. Cicerone, who was project leader, said he began receiving calls perhaps three or four times a week, telling them "to keep quiet about this until a lot more work had been done or to try to downplay it whenever we had to talk about chlorine cycles."

As far as Cicerone was concerned, the most blatant attempts at suppression involved two scientific papers prepared by the Michigan researchers, one for the Kyoto meeting and another submitted to *Science* magazine.

Cicerone and Stolarski found out about the Kyoto meeting from Bob Hudson, who had been invited to give a paper there. During the summer of 1973, he encouraged the Michigan scientists to submit an abstract on their chlorine work. They did so, listing Hudson as a coauthor and omitting any mention of the shuttle. According to Hudson, he was the one who suggested the omission. He was not certain whether the shuttle really was a problem and felt it was sufficient to get the chlorine chemistry out in the open to see if the Michigan team had it right. Consequently, the paper dealt only with natural sources of HCl, primarily volcanos.

Hudson did not realize, until he saw the preliminary program of the Kyoto meeting, that his name had been put on the abstract. This was awkward for him because such things are supposed to be cleared by NASA first. Cicerone felt strongly that Hudson's name should be on the paper. In fact, Cicerone felt more strongly about it than Hudson himself did. Cicerone believes that NASA was pressuring Hudson in its continuing attempt to keep the shuttle problem quiet, but it seems more likely that Hudson was simply disinclined to push the issue of having his name on the paper, perhaps in anticipation of the hassles that might ensue.

It was in part to protect Hudson that Stolarski did not discuss the shuttle at Kyoto. Instead, he gave a brief summary of the work done by the Michigan team, focusing on volcanic eruptions as a source of stratospheric chlorine.

The paper came under strong attack from Mike McElroy of Harvard University. McElroy thought there were serious errors and omissions in the chemistry, and he particularly challenged Stolarski's assertions regarding volcanic sources of chlorine. The Harvard group, he said, had also looked at the effects of volcanos and did not believe they were a major contributor to stratospheric chlorine.

Others in the audience thought the chlorine problem was a complete red herring. They knew that atomic chlorine would destroy ozone, but there didn't seem to be any really significant source of chlorine in the stratosphere. They were unimpressed with the case Stolarski was making for volcanos and wondered why Stolarski and McElroy were wasting everyone's time. Of course, what they did not know was that the two were actually talking about the shuttle. We have seen that Stolarski chose not to enlighten them. Nor did McElroy.

Mike McElroy is one of the most flamboyant personalities associated with the ozone controversy. He is known for the quickness of his mind and his facile ability to assimilate and understand new ideas and information rapidly; it is a talent that disconcerts colleagues and competitors alike. McElroy is extremely competitive and favors a strongly confrontational *modus operandi* in scientific exchanges. In consequence, a series of marked personality clashes has characterized his involvement in the ozone controversy.

McElroy, an Irishman with a shock of red hair and a pale complexion, can be a walking advertisement for the effects of ultraviolet radiation. He once made a memorable impression on Thomas Jukes, a microbiologist who writes a regular column for *Nature* magazine. Jukes wrote of Mike:

> . . . on the beach at Cape Canaveral in Florida, I saw a redheaded man, sunburned to look like a boiled lobster, applying Novocain cream to his glowing back. The only unusual circumstance was that the man was Mike McElroy, whose field is the physics and chemistry of planetary atmospheres and who has loudly warned us against the ultraviolet perils of destroying the ozone layer. . . . Surely he, of all people, should have kept his shirt on.

Unlike Stolarski and Cicerone, McElroy was already directly involved in studying the ozone problem; he was one of those engaged in the CIAP study. Ironically enough, however, it was not his CIAP work that brought McElroy to the shuttle problem; he came to the question of chlorine chemistry by way of Venus. As an expert on planetary atmospheres, he had for many years been involved in NASA's program of exploring other planets with unmanned spacecraft. In 1970, he and his colleagues were studying the chlorine chemistry of the atmosphere on Venus, and their papers on Venus contained the correct chain reaction that was relevant to the ozone problem on earth.

A graduate student had broached the idea of studying chlorine in the earth's lower atmosphere, but McElroy did not think the question was very promising as a research topic, and the matter was not pursued at the time.

According to McElroy, he first began to think seriously about chlorine in the earth's atmosphere in mid-1972, after he saw NASA's environmental-impact statement on the shuttle and recognized it as a potential source of stratospheric chlorine. He and his associate, Steve Wofsy, started working on the chlorine problem, but they did not immediately focus on the shuttle per se. Instead, McElroy said, they concentrated on identifying and studying natural sources of chlorine, including volcanos.

They also calculated how much ozone would be destroyed for a given input of chlorine, without specifying where the input was coming from. This work took them the better part of a year, and it was completed only a few days before McElroy left for the Kyoto meeting.

McElroy was an invited speaker at Kyoto. In his hour-long review of atmospheric photochemistry, he dealt primarily with the role of NO_x. He mentioned chlorine chemistry only passingly, if at all.

Stolarski spoke after McElroy, and it was during the question period following that he and McElroy got into their wrangle over chlorine from volcanic sources. Neither said anything about the shuttle, though they both knew it to be a chlorine source.

Stolarski said that the main reason he kept quiet was because he did not believe that the shuttle's impact on the ozone layer would be large, and he was more concerned about the chlorine chemistry itself. "I didn't think I was talking about the shuttle. I was putting what I considered the scientific end of the paper out front at a scientific meeting."

The Michigan team did not know that McElroy also knew about the shuttle, although Stolarksi said he knew from rumor that Mike was working on chlorine in the lower atmosphere.

McElroy, on the other hand, said he did know at the time of the Kyoto meeting that the Michigan researchers were working on the shuttle. He denies that NASA pressured him to keep quiet about the shuttle; he did not mention it in Kyoto, he said, because there was still considerable uncertainty about how serious the effect would be. It was all "entirely speculative."

The result was that the Kyoto meeting drew to a close with most of the world's top experts in stratospheric chemistry none the wiser about the shuttle's potential impact on the ozone layer.

Stolarski's attempt to spare Bob Hudson embarrassment by not mentioning the shuttle proved to be fruitless. A trip report prepared by another NASA scientist erroneously stated that Hudson had given the chlorine paper and that it had been obvious he was referring to the shuttle. A minor flap ensued, but, fortunately for Hudson, the report temporarily got lost in the maze of the NASA bu-

reaucracy, and by the time it surfaced, he had been able to quieten
people down.

After the Kyoto meeting, Schiff began assembling the material
for the special volume of the *Canadian Journal of Chemistry*. He
spent the better part of the last three months of 1973 chasing after
the invited speakers for the final versions of their papers.

When the Wofsy and McElroy paper arrived in late November,
Schiff was a little taken aback to discover that it was dominated by
a discussion of chlorine chemistry, rather than the nitrogen chemistry that had made up the bulk of McElroy's presentation in Kyoto.
It is perfectly acceptable for scientists to update such papers, but
the changes are normally of a rather minor nature. McElroy explained that he and Wofsy were about to publish the chlorine work
anyway and they had simply decided to put it into the paper they
had to do for the *Canadian Journal of Chemistry* rather than write
a second paper. This revised paper explicitly stated that the impetus
for their chlorine work was the shuttle, but they did not calculate
the shuttle's impact, which they said could not be reliably assessed
until the natural chlorine cycle was better established.

At the end of January, long after the deadline had passed, Schiff
got a rather frantic call from Stolarski asking whether the Michigan
paper could still be included in the published proceedings. This call
had an interesting history. After the Kyoto meeting, Cicerone and
Stolarski had prepared a paper that they submitted to *Science* magazine. This paper, which did not have Hudson's name on it as a coauthor, did mention the shuttle, and Cicerone soon felt himself to
be under considerable pressure from NASA not to publish it. He
said he began to get phone calls from NASA headquarters, some of
which "I construed as threats that our support would be cut off if
we didn't hold off publishing this."

He also remembers a December 1973 meeting at the NASA Ames
Research Center where he and Stolarski were pulled aside by a
group of NASA scientists and "warned that we shouldn't persist in
having our *Science* paper published. They encouraged us to withdraw mention of the shuttle if it should be published. They let us

know they weren't speaking personally, because they were our friends, but they were representing higher interests in NASA."[2]

Stolarski's interpretation of what happened during this period has a slightly different coloration. He said the calls seemed mainly an attempt by NASA officials to find out just exactly what the Michigan group did plan to say publicly. "I think they were worried and they were trying to feel it out. I don't think I would call it overt pressuring."

Nat Cohen of NASA headquarters, who had maintained a continuing interest in the Michigan work, said: "I am not aware of any attempts to suppress that information."

Whatever the case, Cicerone and Stolarski nevertheless submitted their paper to *Science*. It was rejected. One of the reviewers recommended that it be published, but suggested that major technical improvements could be made. The second reviewer, Steve Wofsy of Harvard, recommended rejection on technical grounds, saying the authors had not provided a substantial insight into atmospheric chlorine chemistry.

This was a demoralizing period for Cicerone and Stolarski; the latter, in particular, became rather depressed. This was caused in no small part by the fact that they had received a preprint of the paper Steve Wofsy and Mike McElroy had submitted to the *Canadian Journal of Chemistry*. According to Stolarski, it did not bother him so much that the Harvard paper was mostly about chlorine, though their Kyoto talk had concentrated on NO_x. Updating review papers is done all the time. What made the Michigan researchers "pretty damned mad," he said, was that the paper did not acknowledge that chlorine had been first discussed by Cicerone and Stolarski at the Kyoto meeting. Stolarski believes that McElroy should have recognized the Michigan contribution to the debate—"even if he'd spent two sentences saying our suggestion that volcanos were an important source of chlorine was hogwash."

Stolarski's reaction to this was one of withdrawal. "My initial reaction was, goddamn, let me get out of this deal. This is too much for me." He was not used to the controversy and the professional competition that characterized the stratospheric game. "I had been

---

[2] A NASA scientist who reportedly participated in this discussion said he could not recall it ever having occurred.

sheltered from that type of thing in my previous scientific work," he said. In the ionosphere, there were more scientific problems than there were people to work on them, so everyone was able to give everyone else a wide berth. Such was not the case in the crowded and highly competitive world of stratospheric research.

Cicerone remembers that Stolarski was sufficiently fed up that he was prepared to abandon the attempt to publish the paper. But Cicerone was more inclined to scrap, and he talked Stolarski into reviving the effort to publish in the *Canadian Journal of Chemistry*. Hence the call to Schiff.

Schiff was reluctant to accept the paper because the deadline had long since passed. But he felt some obligation to include Cicerone and Stolarski's work because they had, after all, been the first to raise the issue of stratospheric chlorine. So he agreed to take the paper if the publishers could handle it without delaying the publication of the journal. He also insisted that the Wofsy/McElroy paper take some notice of the Michigan work, so the Harvard paper contained a note added in the proofs that read: "After completion of this work, we learned of a discussion of atmospheric chlorine by Stolarski and Cicerone which was in part presented at the IUGG meeting in Kyoto. Their article also appears in this issue of the *Canadian Journal of Chemistry*."

The Kyoto incident left lasting scars. It generated a profound hostility between the Harvard and Michigan groups that would follow them into the fluorocarbon and fertilizer controversies that were to come.

By January 1974, NASA was taking steps to give the shuttle problem the attention it deserved. The Shuttle Exhaust Effects Panel had decided to sponsor a scientific workshop to be held at the Kennedy Space Center (KSC) in Florida. The Kyoto meeting had demonstrated that there was already a good deal of expertise on the chlorine/ozone problem within the scientific community, and the panel wanted to pick the brains of those who were active in this field of research.

One of NASA's general management review meetings was held early in December 1973. These are regular meetings at which the whole gamut of NASA activities are discussed, and Ron Greenwood briefed NASA Administrator James Fletcher about plans for the upcoming workshop at the KSC.

It is important, in light of subsequent events, to know exactly when Fletcher really became aware of the chlorine problem. Greenwood is not convinced that his briefing really got through to Fletcher because so many things are thrown at the administrator in these meetings. Fletcher says he does not remember the briefing and wonders whether he was even there, since "generally speaking, something like that [the chlorine problem], I would notice."

The three-day workshop at the KSC started on January 21, 1974. In addition to the NASA scientists, there were about a dozen researchers from outside the agency. The discussions were highly technical, and the uncertainties were still too large to say definitely what impact the shuttle would have on the ozone layer. Rough calculations suggested the possibility of a 1 or 2 per cent ozone depletion, but at least one participant suggested that there might be a net increase in ozone.

The report that resulted from this meeting noted that the shuttle would constitute a "small but significant" addition to natural sources of HCl in the stratosphere and said that the calculations of ozone depletion ranged "from significant to insignificant."

Hal Johnston was one of those who attended the KSC meeting, and Hudson remembers that during the early part of the meeting, Johnston seemed to be going through an internal tug-of-war. "It was obvious that Harold was sitting on something that he really didn't want to sit on." Finally, about halfway through the meeting, Johnston began telling people about his discussions with Sherry Rowland concerning the fluorocarbons. The news had not yet gotten around, and Johnston was clearly concerned about the protocol of talking about it, but on the other hand, he wanted to discuss an interesting piece of new research with his colleagues. "It was interesting to see Harold break down and tell us," Hudson said.

On February 1, less than ten days after the KSC meeting NASA's physical sciences advisory committee met at NASA headquarters in Washington. A summary of the Florida workshop was one of the items on the agenda.

In theory, these meetings were, by law, open to the public. However, according to one source with an intimate knowledge of the inner workings of the space agency, NASA did everything possible to thwart the intent of the open-meeting concept and consistently tried to avoid coverage of meetings where sensitive matters that were considered "subject to misinterpetation" were on the agenda.

But NASA was not often able to shake Everly Driscoll, a former Texas high school teacher turned science writer. Driscoll's involvement with the space program had begun in 1969 when she took a job typing up the air-to-ground communications between astronaut crews and mission control for the large contingent of Japanese reporters that descended on the Johnson Space Center for the Apollo moon missions. The Japanese often had trouble following the jargon-riddled dialogue—at times it hardly qualified as English—nor could they wait for the transcripts provided by NASA, which usually ran several hours behind the events. This was not quick enough for the Japanese, who had to write their stories longhand in thin vertical columns of Japanese script, and then file by telephone. They needed someone to type up the air-to-ground communications in simple English as it was happening. Driscoll took the job and was soon a familiar sight, sitting in the Apollo press room, an earmuff-like headset on, typing quickly but calmly, surrounded by the cacophony of dozens of Japanese reporters frenziedly shouting their stories at their editors half a world away.

This was like an immersion course in the space program, and she was later able to switch to covering the moon missions as a reporter when she became space writer for *Science News*. She moved to Washington and later became science writer for the International Press Service of the U. S. Information Agency.

Driscoll regularly attended NASA's physical sciences advisory committee meetings. To her, the February 1 meeting was just one more.

During the morning of February 1, the members of the advisory committee were briefed on the result of the KSC workshop. Ichtiaque Rasool, a high-ranking NASA scientist, discussed calculations of the amount of ozone depletion that the shuttle might cause. It appears that these numbers were in the range of 1 or 2 per cent for a global average, with an effect five times higher in the high-traffic corridor over the Florida launch site. The numbers were quite a bit higher than any that had been previously discussed and, according to one participant, very high figures for the climatic effect of this ozone reduction were also suggested. In fact, the numbers came as a bit of a shock to those at the advisory committee meeting, most of whom had never heard them before. Though the figures were what McElroy would later refer to as "loose numbers," they nevertheless caused no small amount of consternation at the time.

The advisory committee broke for lunch, and when they reassembled after the noon period, Jim Gehrig, a staff member of the Senate Aeronautics and Space Committee, showed up. It was not long before he heard about the morning's discussion of the shuttle/chlorine problem, and he questioned McElroy, who was chairman of the session. According to Tom Donahue, who was standing beside them, McElroy was "bubbling over" about chlorine chemistry. McElroy also put in a plug for the planetary program, saying that the importance of chlorine was recognized as a result of studies of the atmosphere of Venus.[3]

What exactly happened next is a little unclear—remembrances are vague—but it appeared that news about the shuttle traveled quickly to Capitol Hill and discussions ensued at some point during that day between Fletcher and representatives of the Senate Space Committee. Such discussions are alluded to in a letter sent to Fletcher by space committee chairman Frank Moss dated February 1.[4]

Whatever happened, it produced some rather dramatic results, according to Donahue. An hour or so after the afternoon session of the advisory committee meeting got under way, McElroy, who was chairing the session, was handed a note. He looked at it and

[3] This was becoming an increasingly common pitch among planetary scientists as the age of relevance in research overtook the planetary program. The obvious retort of critics is that spending the money on more earthbound projects was even more likely to turn up answers relevant to this planet. In any event, neither the Michigan nor Lockheed teams, which studied the shuttle/chlorine problem nor the Rowland/Molina team, which discovered the fluorocarbon problem, had anything to do with Venus.

[4] The letter requested Fletcher "to provide the Committee with a full and prompt report on the nature of the problems which have been alleged to exist, your views of the validity of these concerns and actions you propose to take, including costs and schedules."

jumped up from the table. Turning to Donahue, he said: "Take over" and hurriedly left the room. The note had apparently contained an urgent summons from Fletcher, and within minutes McElroy was in the administrator's office.

Neither McElroy nor Fletcher will discuss the details of their meeting. Fletcher, in fact, said that he doesn't remember it.

However, shortly after the meeting took place, McElroy related the tale to some scientific colleagues, and according to these accounts, Fletcher was furious. He demanded to know why Mike had allowed discussion of this sensitive issue at an open meeting. In his own defense, Mike reportedly protested that the summary of the KSC meeting had been NASA's agenda item, not his.

In an interview, Fletcher acknowledged that he was concerned by the fact that the shuttle/chlorine problem had been discussed at an open meeting. He recalls the period around late January and early February as the first time he was really aware of the problem, and his concern, he said, stemmed in large part from the fact that he felt he had not been adequately briefed on the problem. "I wanted to know why people hadn't let me know about this earlier because here we have a potential environmental problem, and it didn't get into the environmental-impact statement, and how come I'm just finding out about it now? So I was a little upset with some of our guys."

Ironically enough, those who had been trying to get to Fletcher felt a sense of frustration in battling their way through the many layers of middle management. "There was a vast amount of not taking problems up to Fletcher," said one.

Fletcher himself seems to subscribe to this view: "There are a lot of things like that—that take a long time to get to the top." Yet it doesn't seem entirely fair to blame a nameless, faceless "middle management" for the failures of communication that characterized the shuttle episode. Memos were sent, briefings were held. It would appear that there was really no good reason why Fletcher should have been caught off guard by the events surrounding the advisory committee meeting in early February.

Those events could not have occurred at a worse time. It was budget time at NASA—open season on the shuttle as far as critics were concerned. In a few days, Fletcher had to appear before

the Senate Space Committee to make his annual defense of the
agency's budget of over $3 billion, and there were the usual con-
gressional foes lurking in the wings. The shuttle was particularly
vulnerable. It was NASA's most expensive project; with its annual
costs mounting steadily toward the $1 billion mark, it represented
nearly a third of the space agency's entire budget.

But it was more than just the largest expenditure on the ledgers.
It was also the prestige project, the inheritor of the Apollo legacy.
It was, in fact, all that remained of the man-in-space program.
NASA was not at all interested in a skirmish over the possible effect
of the shuttle on the ozone layer, particularly with the specter of the
now-defunct SST haunting them. Fletcher, who had not been head
of NASA during the SST fight, says he was not worried that the
shuttle would go down the drain like the SST did. But others in
NASA indicate that there was certainly concern—and sometimes
a fear bordering on panic—within the space agency that the shuttle
might well share the SST's unhappy fate if it were to be connected,
however tenuously, with the ozone controversy.

According to one source who was privy to the discussions going
on in NASA at the time, there was less concern about the reaction
of the space committee (which was generally sympathetic to
NASA's goals) to the shuttle/chlorine problem than about press
disclosure of the problem. Though Everly Driscoll was not technically a member of the press at the time—the U. S. Information
Agency is a government agency—her presence at the advisory committee meeting nevertheless caused some consternation. According
to one source, "once the presence of an 'intruder' was confirmed, a
series of hurried high-level meetings was held in NASA headquarters to decide how to deal with the situation."

Driscoll went in to see Fletcher on February 1. She is reluctant to
confirm even that this meeting took place or to discuss what was
said between them. However, she did acknowledge telling Fletcher
that the figures for shuttle depletion of ozone—particularly the corridor effect over Florida—had alarmed her. She said that she urged
Fletcher to put the problem on the public record as soon as possible.

According to other sources, Fletcher asked Driscoll to hold the
story for a while to give him time to assess the situation. She reportedly agreed to hold it as long as NASA was doing something about

the problem, but in any event only unless and until someone else got wind of it. If that were to happen, Fletcher, according to these accounts, promised he would give her an exclusive interview.

Driscoll emphatically denies all of this. "He really did not ask me not to write it. If he had done that . . . that would have made me so angry that the whole story might have been different."

Fletcher's remembrances of the discussion are characteristically vague. He says he was not particularly upset about Driscoll's presence at the advisory committee meeting and he does not think he asked her to hold the story.

On February 5, four days after the advisory committee meeting, Fletcher testified before the Senate Space Committee. He did not volunteer any information about the chlorine problem, nor did the senators question him on it, even though it was clear from Moss's February 1 letter that they, or at least the chairman, already knew about the problem.[5]

It was at this juncture that NASA received a call from Toronto journalist Lydia Dotto, asking to speak to someone in the space agency about the shuttle/chlorine problem. Dotto was in Washington at the time on a previously planned trip related to other stories she was working on. The shuttle had not been on her agenda at all, but just before she left Toronto on February 5, she had talked briefly with Jack McConnell, a York University aeronomer who had attended the Florida meeting. McConnell was unable to say what NASA was doing by way of follow-up to the Florida meeting, so she decided to make some inquiries in Washington. Her call started a chain reaction. How had a reporter from Toronto found out about the Florida meeting? McConnell was the likely culprit. Had he gone back to Toronto and called a press conference? McConnell was also a former colleague of Mike McElroy. So Mike was once again on the spot. "I think there were some people who thought I was somewhere in the loop," he acknowledged.

What exactly transpired at this point is hard to pin down, but the

[5] According to an internal NASA memorandum written at the time, Fletcher met with Moss around this time and "Moss evidently told [Fletcher] not to open up to the committee on this in the next day's hearings."

result was rather dramatic. McElroy phoned McConnell in a state of great agitation, demanding to know what he had told the press. Apparently not realizing that Dotto was already in Washington, McElroy wanted to know if there was some way of heading her off, wrongly construing her trip as a direct assault on the shuttle.[6]

McConnell, a gentle soul, was nonplused by the fierceness of McElroy's attack. McConnell admitted that he had talked to Dotto, but defended himself by saying that no one at the KSC meeting had said anything about it being secret.

McElroy painted dire pictures of the imminent collapse of the shuttle program and said that NASA had opened "files" on McConnell and Dotto. (Interestingly enough, McElroy's *Canadian Journal of Chemistry* paper was on the verge of becoming the first paper in the open scientific literature to name the shuttle as a source of stratospheric chlorine.) Although something clearly ignited McElroy, it would appear that the intensity of his outburst stemmed more from his own gift for emotional hyperbole than from a NASA edict. No one in NASA seemed to know anything about the alleged files, and it seems doubtful that they existed. Fletcher's reaction to this episode was one of disbelief; told of McElroy's call to McConnell, he said the story must be apochryphal. "Oh you must be kidding. I don't ever do stuff like that." Asked if NASA had opened files on McConnell and Dotto, he said: "Over my dead body."

On February 14, Bob Hudson and Ron Greenwood briefed Fletcher on what had transpired at the Florida meeting. They talked to the administrator for about forty-five minutes, and Hudson got the impression that this was the first detailed scientific explanation of the chlorine problem that Fletcher had received.

Fletcher, however, says he does not remember this briefing. He recalls hearing about the Florida meeting from Ichtiaque Rasool, although Fletcher is uncertain whether this was before or after the

[6] Meanwhile, in Washington, Dotto was speaking with Homer Newell, one of NASA's associate administrators. He gave her a copy of the 1972 environmental-impact statement, whose only relevance to the chlorine/ozone issue was its lack of attention to the subject. As to the chlorine problem, Newell's pitch was composed in equal parts of NASA party line and bland assurances that the agency did not believe the shuttle would be a problem. But in fact, at that point, the only thing that could legitimately be said is that no one knew whether the shuttle would be a problem or not.

advisory committee meeting on February 1. He also says he does not remember either Greenwood or Hudson, two scientists who were, at the time Fletcher made this statement, in charge of NASA's stratospheric research program.

Hudson remembers that Fletcher did not seem unduly upset about the shuttle problem during the briefing, but the administrator readily agreed that there must be a program to study the problem. Shortly thereafter, a space shuttle environmental directorate was established at headquarters and an environmental-effects project office was set up at the Johnson Space Center under Bob Hudson.

Fletcher went public with the chlorine problem the day after he was briefed by Greenwood and Hudson. The occasion was a meeting of the National Space Club in Washington, where Fletcher was guest speaker. His speech was a lengthy one—a grand tour of highlights of the entire space program. Discussion of the shuttle took up only part of it, and most of that was devoted to economics. Reporting on the probable outcome of that year's budget battle, Fletcher optimistically told his audience that "we got the space shuttle over its last big hump last week."

The speech contained a few paragraphs on the shuttle's environmental impact, the last of which read:

. . . our continuing studies have shown that the hydrogen chloride in the booster exhaust may give rise to free chlorine in the stratosphere, which laboratory experiments indicate might catalyze the destruction of some stratospheric ozone. There are, however, no data to show that this actually happens and it is an extremely complicated question that we will continue to study. We fully expect that this will not turn out to be a problem but should the effects turn out to be unacceptable, there are alternative propellants that we can use in the rocket booster and we will do so.[7]

[7] Keep in mind that NASA was already committed to the $106 million contract for the rocket boosters to Thiokol. Fletcher says that quick calculations were done just prior to his February 14 speech and NASA concluded that the economic penalty of getting out of the contract would not be "severe." NASA later estimated that the development cost for a new propellant would be around $80 million.

This action of Fletcher's surprised some people within NASA, but his comments attracted little attention outside the space agency and virtually no media coverage.

Also on February 14, Fletcher responded to Moss's letter requesting information on the shuttle/chlorine problem. Fletcher said the matter was under study, but there was, as yet, no real proof that a problem even existed. He did not expect that NASA's studies would show the shuttle to be a danger to the environment, but he repeated his pledge that the rocket propellants would be changed if necessary. "The shuttle will be designed to have no harmful effects on the atmosphere, either in the troposphere or stratosphere." Both letters were made public during Senate hearings in late February.

In July 1977, NASA released a revised draft environmental impact statement for the shuttle. It stated that 60 shuttle launches a year would result in ozone reductions in the Northern Hemisphere of about .2 per cent, with an uncertainty of a factor of 3 in either direction. (The National Academy settled on a figure of .15 per cent ozone reduction, with a factor of 3 uncertainty. In mid-1977, the NASA figures were upped by a factor of 2 as a result of new data.) The NASA environmental impact statement also noted that the effect in the Southern Hemisphere would be smaller, by a factor of 5 to 10. According to the impact statement, the maximum impact on ozone would occur just a few years after the shuttle started injecting chlorine into the stratosphere (unlike the case of fluorocarbons, where the full impact would be delayed for a few decades). Moreover, NASA calculated that it would take only a few years for the ozone layer to recover if the shuttle booster was changed to one that did not emit chlorine. (If the boosters were changed in 1992, for example, the ozone layer would return by the year 2000 to the levels that existed in the early 1980s).

The shuttle incident was an important milestone in the ozone controversy. Just as the SST had forced scientists to consider the role of man-made NO_x in the earth's upper atmosphere, so the shuttle forced them for the first time to look at the stratospheric effect of chlorine compounds from human activities. Until the shuttle came along, the ozone-destroying capability of chlorine had been largely a textbook or laboratory curiosity.

The shuttle problem had another important effect: It injected NASA into the ozone debate in a major way. The space agency made a big play to become the "lead agency" heading the government's stratospheric research effort. Ironically enough, NASA could easily have beaten everyone else in these sweepstakes had it funded the proposal to study fluorocarbons that Charles Kolb had submitted in the fall of 1973.[8] But what the space agency may have lacked in foresight it more than made up in aggressiveness as it sought to convince Congress that it would be the best lead agency to run the stratospheric show.

This inspired no small amount of cynical comment. NASA had fallen on hard times since the glory days of the Apollo moon landings, and some critics saw, in many of the agency's subsequent activities, a rather desperate attempt to come to terms with the social and political climate that demanded down-to-earth relevance from NASA. But there were those who could not see an agency devoted to aeronautics and space flight as the logical choice to investigate the chemical effects of spray-can propellants on the upper atmosphere. Nor did it escape attention that, with its involvement in the shuttle, NASA had a certain vested interest in the outcome of such studies.

However, Tom Donahue said that it was he and Mike McElroy who sold NASA on the idea of going after the stratospheric research program. They had come to the conclusion that the space agency was best equipped to study the whole ozone problem, not just the fluorocarbon issue, and put this suggestion to George Low, Fletcher's deputy. They urged NASA to "take the bull by the horns." Shortly thereafter, NASA did just that.

The role of the shuttle is probably the least-known aspect of the ozone controversy. If you ask the proverbial man-on-the-street about threats to the ozone layer, he will probably mention spray

---

[8] It is also ironical that Mike McElroy, who would later figure so prominently in the fluorocarbon debate, missed this chance. Kolb told him about the fluorocarbons one evening in the fall of 1973 when the two of them happened to bump into each other at a restaurant. Kolb was anxious to have McElroy work with him on the problem but "we just never connected," said McElroy. "We got off on different tracks. There seemed to be more important things to do at the time."

cans, he may even mention SSTs, but the odds are high that he will not mention the shuttle.

In part this is because NASA, while perhaps not engaged in out-right suppression, did rather successfully stage-manage a low-profile treatment of the shuttle/chlorine problem. More importantly, how-ever, the shuttle problem was rapidly overtaken by the spray-can controversy. Fletcher went public with the shuttle problem in mid-February 1974; by June, the Rowland/Molina paper on fluorocar-bons was published, and by fall, the spray-can issue had moved front and center.

There could be no low profile for the spray-can controversy. Whereas the shuttle and the SST were esoteric technologies that would never directly touch the lives of most Americans, the spray can was a ubiquitous technology familiar to all Americans. Whereas NASA had managed to keep the shuttle out of the headlines, the aerosol industry could not hope to do the same for spray cans; their only option was to meet the challenge head-on. Thus began the elaborate public-relations and advertising campaign to save the spray can.

CHAPTER SEVEN

The Sky Is Falling

We beg to differ with Chicken Little.

—Aerosol industry press release.

On February 1, 1975, the spray-can controversy made a cameo appearance on the television show "All in the Family." Mike and Gloria Stivic are having an argument over whether or not to have a child. Gloria wants one; Mike doesn't because the world is in such terrible shape. To illustrate his point, he picks up a can of Gloria's hair spray and tells her that it will destroy the ozone layer.

The fact that the spray-can controversy surfaced on "All in the Family" had to be an ominous portent to the aerosol and fluorocarbon industry. Publicity like this they did not need. "All in the Family" reached some forty-three million people a week in early 1975, and the popular show was noted for its attention to current, controversial subjects of wide public interest. The Rowland/Molina theory that the spray-can propellants were destroying the ozone layer had clearly made the grade. It was not going to go quietly away.

The Spray-can War reached its peak of intensity in 1975 and the first part of 1976. The aerosol industry, government agencies, and the scientific community became increasingly embroiled. The congressional mail generated was reportedly greater than that for any issue since the Vietnam war. Ultimately, the spray-can threat to the ozone layer would be confirmed, but not before an intensive two-year research effort had been completed and not until dramatic battles had taken place on three fronts: the public-relations front, the political front, and the scientific front. Of course, these battles were intimately related to each other, and they occurred in parallel. But each had some distinctive features, and it is useful to consider them separately. We will look first at the public-relations battle. As the one that most directly involved the consumer, it was the most visible and, in many ways, the most entertaining.

However, to understand this battle and the others, it is necessary first to understand just what it was that the fluorocarbon and spray-can companies were fighting for. The aerosol industry had enjoyed a phenomenal growth in the United States. In the late 1940s, less than 5 million cans were sold each year. A decade later, annual sales had passed the 500 million mark. Between the late 1950s and the early 1970s, sales increased nearly 500 per cent. They peaked in 1973, just before the word "ozone" entered the picture and started making life miserable for the aerosol industry, when 2.9 billion spray cans were filled in the United States alone (about half the world total). The average American household contained an estimated 40 to 50 cans of various sorts—shaving lathers, hair sprays, insecticides and insect repellants, waxes and polishes, cooking sprays, pharmaceuticals, deodorants and antiperspirants, and room disinfectants. By the mid-1970s, some 300 products were packaged as aerosols. About half the aerosols produced each year contained fluorocarbon propellants. These were mostly hair sprays and deodorants. Many other products used different propellants, such as hydrocarbons and nitrous oxide (or mixtures of these substances, sometimes including small amounts of fluorocarbons).

In 1973, about half of all fluorocarbons produced went into spray cans, mostly into hair spray and deodorant containers. They have several advantages as aerosol propellants. They are odorless, non-toxic, and chemically inert. They do not spray on cold to the touch (an advantage in deodorants), and they provide a finer mist than other propellants do, which ensures that hair sprays did not go on in a thick, gluey mass. Another advantage of fluorocarbons, particularly for personal products, is that they are nonflammable. Unlike most other spray products, hair sprays and deodorants are directed toward the person using them. If that person also happens to be smoking, disastrous consequences might result if the propellant was flammable. A hairdo going up in flames is not an appealing image.

Before the fluorocarbon controversy started, many consumers probably did not realize that the propellant made up a large percentage of the contents of a spray can. Fluorocarbons made up about 65 per cent by weight of the contents of an average can of hair spray and about 90 per cent of a can of deodorant. Less than 1 per cent of the weight of the contents of the can was the active ingredient.

Slightly more than a quarter of the fluorocarbons produced went into air conditioners, home refrigerators, and commercial-industrial refrigerators and freezers. The remaining fluorocarbons were used for a variety of other purposes, including foam-blowing to produce such products as styrofoam cups and insulating material for buildings.

Spray cans posed the major threat to the ozone layer because all of their propellant is deliberately released and quickly gets into the atmosphere. In 1973, spray cans accounted for about 75 per cent of all emissions of fluorocarbons in the United States.

About 15 per cent of the emissions came from refrigeration and air-conditioning systems. Domestic refrigerators and air conditioners were not the main culprits; they are hermetically sealed and often retain their fluorocarbons for the lifetime of the appliance; car air conditioners, however, were notoriously leaky.

Most of the refrigerant emissions came from commercial refrigeration units—meat-packing plants, vegetable-storage systems, machines that make freeze-dried coffee—and from air-conditioning units that cool large commercial buildings. These systems were not hermetically sealed, and their supply of fluorocarbons often needed replenishing.

The worldwide production of fluorocarbons reached .8 million tons in 1974, about half of it manufactured in the United States. Between 1964 and 1974 the use of fluorocarbons increased at a rate of nearly 10 per cent a year.

By the early part of this decade, fluorocarbons were a flourishing business; when the ozone controversy broke, the industry could come up with an imposing set of economic figures to impress the public and political leaders at a time of recession and high unemployment. By 1973, according to the industry's calculations, the spray-can business in the United States dependent on fluorocarbons was worth nearly $3 billion. Refrigeration and air-conditioning systems, which use fluorocarbons as coolants almost exclusively, were a $5.5 billion industry. Estimates of the employment related to fluorocarbons ranged from 200,000 to 1 million.[1]

But such statistics should be viewed with caution. These figures included the cost of the containers—not only spray cans, but also refrigerators, freezers, and air conditioners. The value of the fluorocarbons themselves was much less. The employment figures are also a bit of a numbers game. An industry-sponsored study showed that only some 2,000 people were employed directly by the fluorocarbon propellant producers. If you counted the employees of the container companies, the valve manufacturers, the cap and label suppliers, the companies that filled aerosol cans, and the companies whose products were sold in the cans, the total still came to just below 40,000. The figure of 200,000 or more put out for public consumption must have included not only those who produced the cans and the refrigerating appliances, but also probably those who sold them as well and anyone else who had anything to do with them. As one cynic observed in *Rolling Stone* magazine, "they must be down to counting air-conditioner repairmen by now."

Nevertheless, inflated figures notwithstanding, the economic stakes were high. This was a major reason why the industry fought so hard to save the fluorocarbons.

There is a temptation to think of the aerosol/fluorocarbon industry as a monolith, as a single entity with a single goal. But this was not the case. Not only were there many different companies involved in the Spray-can War, there also were many different *kinds* of companies.

To the public, the most visible ones were the marketers. These were the companies that actually made the product—hair spray, deodorant, or whatever—that went into the can. These were the companies whose brand name went on the cans.

---

[1] A 1976 study by the federal Commerce Department said that 600,000 jobs were directly dependent on the production and use of fluorocarbons, and an additional 1.5 million jobs were indirectly dependent.

At the other end of the scale were the fluorocarbon manufacturers. There were six in the United States: Du Pont, Allied Chemical Corporation, Union Carbide Corporation, Penwalt Chemical Corporation, Kaiser Aluminum and Chemical Company, and Racon, Inc. Du Pont dominated the field as the largest single producer in the world and manufacturer of half of the fluorocarbons used in the United States; not surprisingly, it also dominated industry's fight to save the chemicals.[2]

Between the marketers and the fluorocarbon producers there were a number of other companies that supplied components such as the cans themselves, valves, caps, and labels.

Finally, there were the fillers. These companies worked under contract to the marketers, who did not themselves actually fill their own spray cans. The fillers never made or sold any aerosol products. They would buy the cans and the fluorocarbons, put the propellants and the product (hair spray, perfume, deodorant, etc.) into the cans, and ship the completed package back to the marketers, who would then distribute the cans under their brand name. The fillers made up a large part of the aerosol industry, accounting for about a quarter of the employment.

Du Pont once estimated that some 53,000 firms in the United States were involved in the production of fluorocarbon-using spray cans.

This polyglot collection of companies with an interest in fluorocarbons cemented their relationships through a number of trade associations and organizations. The two most established of these were the Chemical Specialties Manufacturers Association (CSMA) and the Manufacturing Chemists Association (MCA). Neither of these represented the aerosol or fluorocarbon industries exclusively. However, the CSMA did have an aerosol division and, in response to the ozone issue, MCA formed a fluorocarbon technical panel, which took on the task of despensing $3 million to $5 million of the industry's money for research on the problem.

[2] Du Pont had a particularly large stake in the outcome of the ozone debate. In late 1974, it had just completed construction on a brand-new $100 million fluorocarbon plant in Texas, the largest in the world. A subsequent Commerce Department study indicated that the company had a $250 million annual market for fluorocarbons in the United States alone, of which $50 million was for aerosols.

Of more recent vintage were the Aerosol Education Bureau (AEB)—the industry's public-relations and propaganda arm—and the Council on Atmospheric Sciences (COAS). Both of these organizations concerned themselves almost exclusively with the defense of fluorocarbons and spray cans. COAS in particular was formed early in 1975 to provide paratroops to fight the Ozone War. Operating with a budget of several hundreds of thousands of dollars, it organized industry spokesmen to testify at federal and state hearings and also put industry scientists and spokesmen on the public-speaking circuit.[3]

Through these organizations, industry tried to present a united front to the world. But there were divergent interests and, as we shall soon see, there were internecine struggles that could not be kept from public view.

Finally, there were the trade magazines that chronicled the Spray-can War from industry's point of view. Two in particular—*Aerosol Age* and *Drug and Cosmetic Industry*—did so in exhaustive detail. *Aerosol Age* was nonpareil, unexcelled in the humorless stridency with which it denounced all who had anything bad to say about spray cans. *Drug and Cosmetic Industry*, on the other hand, had Donald Davis, an outspoken writer who said exactly what was on his mind whether the other members of the aerosol community liked it or not. Though he was unquestionably pro spray can—he had been writing about them for twenty years—Davis was often bluntly critical of industry's handling of the ozone problem, which he regarded as the most crucial issue in the history of aerosols. Nor did he wear rose-tinted glasses, as editors of other trade magazines often seemed to do; in fact, he once publicly accused the other magazines of "whistling in the dark" and feeding their readers "overly reassuring words" on the ozone crisis.

Industry began mobilizing for the Spray-can War early. There were some who dismissed the ozone controversy and the attendant anti-aerosol sentiment as just another fad of the eco-freak fringe, but for the most part, industry took the threat seriously. As early as

[3] Marketers were asked to contribute $.50 per 1,000 aerosols sold; components suppliers, $2.00 per $1,000 worth of components sold; contract fillers, $.10 per 1,000 aerosols filled.

the fall of 1974, the Chemical Specialties Manufacturers Association sponsored a seminar on the ozone problem—perhaps the first major attempt by industry to come to grips with the issue that was beginning to engulf them. "If you feel surprised, bewildered, and apprehensive about this, you are not alone," said Igor Sobolev of Kaiser Aluminum and Chemical Company, one of the fluorocarbon producers.

It was clear at the outset from the very nature of the Spray-can War that public relations were going to be important. Aside from the fact that the consumer had to be wooed in the face of this new environmental threat, there were the media to contend with. If the media were slow off the mark, they had, by the end of 1974, gone for the spray-can story in a big way, almost as if to make up for lost time. The reporting at this stage leaned toward the lurid. "Is the homely aerosol spray can and its charge of propellant gas sowing the seeds of doomsday, threatening to destroy earth's ozone shield and bake the planet barren with solar radiation?" asked the Associated Press.

"Remember that each one of us can now 'push the button' on our own," said *Harper's Weekly,* invoking a nuclear analogy. "Listen for that little whisper of doom."

Another story confidently announced that "aerosols have probably doomed more people than were killed by the atomic bomb dropped on Hiroshima." This particular article, in *New Times* magazine, painted grim pictures of new ice ages and human beings becoming a nocturnal species to protect their eyes. "Farmers would have to plow what fields survived at night, [sic] while glaciers thousands of feet thick, leveled cities across Europe and North America."[4]

For sheer melodrama, however, none equaled a Philadelphia

---

[4] This article appeared on March 7, 1975. *New Times* would later claim in an ad that it had "sounded the first national alarm on the little aerosol cans that could be the death of us all"—a claim that was hardly justified in light of the fact that the story had already been covered extensively by the New York *Times,* the Los Angeles *Times,* the *Christian Science Monitor, Time, Newsweek,* and *Business Week,* plus national television and several widely read science magazines. Moreover, an environmental group had already petitioned the government to ban spray cans, and a bill to that effect was under consideration in Congress. The aerosol industry was not the only player in the Spray-can War capable of a little PR hype.

Inquirer story by Ron Chernow that painted a picture of Rowland and Molina bearing an "Atlas-like burden." The article proclaimed that "the world will end—not with a bang, not with a whimper, but with a quiet pfft. . . . The earth may already have committed partial suicide or at least severe self-mutilation." Scientists had "taken this apocalyptic vision and translated it into chillingly specific statistics."

Another news story received hoots of derision from industry for its inadvertent invention of a new aerosol product—"underarm insecticides."

However, the Oscar for Special Achievement in Sensationalizing the Ozone Controversy must go to the movie *Day of the Animals*, released in early 1977. In the show, a group of hikers is savagely attacked by a horde of crazed vultures, rats, dogs, wolves, rattlesnakes, mountain lions, and bears. The reason they are running amok, it is explained, is that the ozone layer has been damaged and the ultraviolet radiation is driving them mad.

Sherry Rowland was asked to attend the preview of the movie. He declined.

Many in industry were irritated—and sometimes justifiably so— by media exaggeration and the tendency to emphasize the cataclysmic consequences of total ozone destruction. But if they were irked by such reporting, they were outraged by articles that portrayed industry representatives as ogres and unprincipled villains. One story in the New Haven *Register* raised hackles when it said of one industry spokesman that "critics maintain that beneath that disguise of Kris Kringle affability lurks one of the men who may cause the end of our present-day earth."

A consumer who had expressed concern about the ozone layer received the following letter (reprinted in *Aerosol Age*) from Montfort Johnsen, vice president of Peterson/Puritan, an aerosol filler: "In contrast to the image often presented by the press, your fellow Americans in the business community are not 'unfeeling monstertypes' but sensible responsible men and women just like yourself. We get terribly concerned when allegations are raised that we are unwittingly tampering with the destiny of our own lives and those of our children. We also get very frustrated because only the most negative of the possibilities are ever broadcast by the press and other media. We have no real way to tell what we feel is the true and balanced story."

In this last comment, Johnsen touched on a common theme among industry spokesmen—their resentment toward the media. The media were generally portrayed by industry in one of two ways. The first image was that of an active enemy interested only in scare headlines that would sell newspapers or grab TV audiences.

For example, John Ferring, president of Plaze, Inc., a St. Louis aerosol filler, said: "I believe the newspaper has dangerously maneuvered this unproven theory into a factual statement. . . . It's tough to fight these mammoth headlines and articles that the media has published as they control the space and maneuver at will."

A second and somewhat contradictory image of the media favored by some in the aerosol industry was that of helpless dupes manipulated by scheming environmental and consumer pressure groups and anti-aerosol scientists.

In *Harper's Weekly*, Robert Abplanalp, president of Precision Valve Corporation, accused Rowland and Molina of having "chosen to run to the media." Abplanalp, a friend and benefactor of former President Richard Nixon, had made millions with his invention of the first inexpensive workable aerosol valve, and his company manufactured about half the world's supply.[5] He was, not surprisingly, outspoken in his contempt for the ozone theory of Rowland and Molina, and said that reporters did not have the knowledge to check the scientific claims. "They merely wrote what they were told and perhaps, as witlessly as Henny Penny, proceeded to inform everyone that the sky is falling."

It was, of course, not true that reporters simply "wrote what they were told" by Rowland and Molina or anyone else. Nor did they exclusively interview only those scientists who had come out against spray cans. Many reporters consistently checked their stories with a large number of scientists who were doing research in this field or were experts in stratospheric chemistry or other important aspects

[5] Abplanalp reportedly received hate mail stained the color of blood from members of the Charles Manson gang. Members of the Manson gang have from time to time publicly threatened to visit retribution upon executives of companies they feel are harming the environment.

of the problem. The list included researchers considered by the industry to be sympathetic to the industry's cause (although the scientists themselves might not put it that way), and others who had no vested interest in the fluorocarbon controversy but who had expertise in certain specific aspects of the problem. Reporters would periodically "do the rounds"—which usually included a call to an industry scientist. In general, the media did not, as industry spokesmen often claimed, consistently fail to give industry's side of the story. And although some of the reporting in the early part of 1975 did get rather overblown and prophesies of the apocalypse continued to pop up here and there, a large proportion of the news coverage in the latter part of 1975 and during 1976 was balanced and responsible. It just didn't say what the industry wanted it to say. Some members of the industry, in consequence, developed acute symptoms of the "shoot the messenger" syndrome.

In its own trade press, the aerosol industry was not infrequently portrayed as a whipping boy—the poor, downtrodden victim of environmental vigilantism at the hands of publicity-seeking scientists and power-hungry regulators in league with the sensational media. Consumer groups were portrayed as all-powerful.

There was also concern within the aerosol industry that the public opposition to spray cans received much of its impetus from the anti-industry, anti-everything mood that seemed to have swept the country. It was perhaps inevitable that someone would see a Communist plot in the anti-aerosol campaign. George Diamond, president of Diamond Aerosol Corporation, once suggested that a computer simulation of the ozone issue would indicate that scientists who mislead the public are fools and traitors, that free enterprise and democracy were being attacked, and that hysterical politicians were behaving suspiciously. Proposed remedies could include investigation of the politics and "suspicious connections" of those involved. (Some time later, Diamond suggested that the Soviet KGB may have been involved in the fluorocarbon controversy.)

The image of the billion-dollar fluorocarbon and aerosol industry —containing some of the largest chemical companies in the world— as a poor, put-upon, helpless victim was a little difficult for many people to swallow, particularly when one considers the sheer money and manpower thrown into the job of defending spray cans. Although industry ultimately developed a sophisticated and high-powered PR program, the ozone problem initially caught them somewhat off balance. There was a period early in the controversy, in late 1974 and early 1975, when their defenses were in a state of disarray. Malcolm Jensen, president of the Can Manufacturers Institute, was reported in *Drug and Cosmetic Industry* as saying that the industry had done a "lousy job" of selling itself. It was "too damned afraid" to make a good case for itself. And the redoubtable Donald Davis wrote that "a long series of public relations disasters has beset the aerosol industry to a point where it seemed that every corrective step, every defensive reaction, every affirmative declaration was made either too late after the fact or with a peculiar lack of conviction obvious enough to make the gesture self-defeating. . . . The industry has been staggering from disaster to disaster, its public image in tatters."

Davis had barely written this when industry's Council on Atmospheric Sciences pulled off a PR venture that backfired rather badly. COAS was rapidly becoming a laughingstock within the scientific community. Despite the authoritative and official-sounding ring to its name—surely no accident—COAS was not a council of scientists. Even industry scientists would greet questions about COAS with loud guffaws. One suggested that its name should have been CHAOS. Another, who nevertheless publicly participated in some of COAS's activities, said its name was "the misnomer of the year. There are no scientists. There is a lot of hot air." Privately, industry scientists would dismiss COAS as an organization run by the marketing and advertising people. Du Pont scientists were among those who were privately most critical of COAS, but the company was nevertheless active within the organization. A. H. Lawrence, director of sales for Du Pont's Freon Products Division, was vice chairman of COAS and actively exhorted others in the industry to contribute to the war chest that COAS was using to defend fluorocarbons. Du Pont scientists were available to participate in COAS-sponsored press conferences and tours, and a Du Pont publication put out by the Freon Products Division wrote approvingly of many COAS activities.

In one of its first major PR offensives, COAS sponsored a six-week visit to the United States by British scientist Richard Scorer. Scorer initiated the contact with the U.S. industry, according to Ralph Engel of the Chemical Specialities Manufacturers Association, who said Scorer contacted them and asked to present his views to the American public. Scorer was a professor of theoretical mechanics at the Imperial College of Science and Technology in London. He had done research on air pollution and was well known in his own field, but he was not an expert on stratospheric chemistry. He had done no research on the subject and had published no scientific papers on it.

Scorer had been active in Britain on the SST/ozone controversy and had, in fact, described the CIAP study as "pompous claptrap." Of the SST controversy, Scorer wrote in the *New Scientist:* "How could it be alleged seriously that the atmosphere would be upset by introducing a small quantity of the most commonly and easily formed compounds of the two elements [oxygen and nitrogen] which comprise 99 per cent of it?" This comment demonstrated an apparent ignorance of chemistry. For one thing, Scorer made no mention of the fact that these "easily formed compounds" would be intolerable for humans if they were formed in the atmosphere in appreciable amounts. One, nitrous oxide, is an anesthetic and another, nitrogen dioxide, is a highly poisonous gas. More importantly, it ignores the fact that "a small quantity" of a chemical can chew up a lot of ozone because of the devastating effectiveness of catalytic chains.

Scorer's remark also demonstrated once again the tendency for atmospheric scientists to simply dismiss chemistry as unimportant. (He referred to meteorologists as "more mature scientists" than those who had proposed the NO_x problem. Ironically, both of those primarily responsible for proposing the problem—Paul Crutzen and Hal Johnston—had initially been trained as meteorologists.)

Scorer evinced the typical British disdain for the furor over spray cans in the United States and, upon arrival here, promptly and emphatically declared the ozone/fluorocarbon theory to be "utter nonsense."

Scorer traveled from New York to Los Angeles carrying the message that the earth's atmosphere is "the most robust and dynamic element in the environment" and that "man's activities have very little impact on it." (He apparently did not notice the losing battle the atmosphere was waging against human activity in Los Angeles.) Scorer said that the atmosphere was far too complex for the simple-minded theories about ozone destruction to be correct. He evidently did not find this to be inconsistent with his own rather categorical statements that all was well with the atmosphere.

Industry representatives professed themselves pleased with Scorer's assault on the fluorocarbon/ozone theory. Ralph Engel of the Chemical Specialties Manufacturers Association said that sur-veys taken before and after Scorer's visit reflected a 50 per cent in-crease in the public's awareness of the other side of the story.

While Scorer may have had some impact on the media and the public, however, his campaign was largely ineffectual on the scientific front. His pitch really consisted of little more than bland assurances about the resiliency of the earth's atmosphere. He said nothing that the industry's scientists had not already said, and he contributed little of substance to the debate. Bringing him to the United States had been an out-and-out PR tactic, something that even some industry officials admitted privately. Industry scientists seemed to blame the episode on the marketers.

Du Pont privately disowned Scorer almost immediately. The tour had barely begun when Ted Cairns, Du Pont's research and devel-opment director, called the National Academy of Sciences, which had convened a panel of experts to study the flourocarbon problem. According to a memo sent to panel members, Cairns, who was a member of the Academy, "wanted the panel to know that he played no part in bringing Scorer to the country and, in fact, had tried to discourage the visit. He does not feel that Scorer has anything new to contribute to the question and feels that Scorer is only likely to polarize the community, something that Cairns wants very much to avoid; Cairns believes that a rational assessment of the nature and extent of the problem is of paramount importance, that steps are being taken in this direction but Scorer's trip will not advance this."

But it was George Alexander, the science writer for the Los An-geles *Times,* who publicly announced that the Emperor had no clothes by characterizing Scorer as a "scientific hired gun . . . Last week, like the nineteenth-century railroad barons who were not above importing a hired gunslinger to deal with a troublesome situation, the chemical industry brought in Dr. Richard Scorer from England to shoot down Rowland, Molina, and the others who have been making life difficult."

Scorer did a lot of shooting, but he didn't hit anyone. In Los Angeles, practically in Sherry Rowland's backyard, Scorer denounced scientists, unnamed, who promoted "scare stories" based on limited scientific evidence. Rowland, however, declined to engage Scorer in verbal combat; he regarded the British scientist as a polemicist with little to contribute to the debate. "The gentleman is good at attacking, but he has never published any scientific papers on the subject," he told one California paper.

However, Mike McElroy took Scorer on. It wasn't exactly Main Street at high noon, but the two appeared on William Buckley's television show, called, appropriately enough, "Firing Line." Scorer is no shrinking violet, but McElroy is a skilled performer, comfortable and effective on TV, and he walked all over the British scientist.

In all, Scorer's tour was of dubious value to the aerosol industry. Perhaps the most intriguing assessment of his visit was made by an industry source, Donald Davis, editor of *Drug and Cosmetic Industry,* who described Scorer's trip as a "road show." He expressed the opinion that "Scorer's rhetorical skills were interesting but only moderately effective. . . . He did a good service from a public-relations standpoint, but relatively little to shake the essential Rowland/Molina theory, especially with the scientific community. . . ." Davis concluded that there may have been some benefit in exposing local media in various cities to a scientist who defended aerosols,[6] but the gains were offset by the "understandably cynical criticism" of Scorer's role.

COAS later did acquire a scientific adviser, James Lodge. Lodge had been at the National Center for Atmospheric Research in Boulder, and was, at the time of his appointment to COAS, chairman of the Colorado Air Pollution Control Commission. His involvement with COAS became controversial, for there were those who deemed his dual role as an industry spokesman and a pollution-control

[6] One industry PR man was quoted in the San Francisco *Examiner* as saying that there were "not too many scientists willing to back the industry's position."

official to be a conflict of interest. Lodge did not agree, but he resigned his Colorado position, saying that he would need the time to devote to his COAS activities.

Lodge was featured prominently in industry press releases, one of which figured in a showdown that industry representatives had with Dorothy Smith, news manager for the American Chemical Society, during an ACS meeting in San Francisco. An industry official she did not know entered the press room and put some news releases on the table she had set aside for press information. When she ambled over to see what this was all about, she discovered a statement attributed to Lodge. It was standard industry cant. Smith recalled that it essentially said that "nothing much had been proved." But what made her angry was that it said Lodge had made this pronouncement at the ACS meeting. Lodge was not presenting a paper at the San Francisco meeting, nor had he been invited to participate in any symposium or press conference so, as far as Smith was concerned, he had no right to associate the ACS's name with the statement. She turned to one industry representative standing nearby and told him she was not going to put up with this. At an upcoming press conference on the fluorocarbon issue, she was going to announce that Jim Lodge's statement had nothing to do with the ACS, "that whatever he was saying, it was not connected with us." The man she was talking to asked if she would reconsider if they agreed to strike the name of the ACS from the news release. She said yes—as long as Lodge did not attempt to make his statement at the press conference. "The moment he opens his mouth, I plan to make that announcement," she said. It was not that Smith didn't want industry's viewpoint represented at the press conference. Peter Jesson, a Du Pont scientist who had been an active industry spokesman, was scheduled to appear, and Smith felt that she had a balanced panel for the press to interrogate. She just didn't want things to get out of hand.

These difficulties notwithstanding, the industry eventually evolved a PR strategy that was slick, expensive, and sophisticated. Unlike their counterparts in the SST controversy, they were used to dealing with the public. Arnold Goldburg, who, as chief SST scientist for Boeing, had had a taste of the public-relations battle himself, felt that the aerosol industry handled its defense much more effectively than did the aviation industry. But then, no one ever sold SSTs in the supermarket.

(Some of the aviation industry's press releases were relatively amateurish efforts full of scientific errors. For example, one put out by the British makers of Concorde, in trying to make the point that nature adds more NO_x to the atmosphere than would SSTs, listed earthquakes as a natural source of NO_x. They presumably meant volcanos, which are, in any event, a negligible source.)

In private, fluorocarbon industry scientists often endeavored to put some distance between themselves and the PR operation, which, they said, was dominated by their companies' marketing and advertising people. Ray McCarthy of Du Pont said that one of the most difficult problems faced by industry scientists was convincing industry PR people that "something that was not proved was not automatically untrue."

Much of the industry's PR campaign was handled by the Aerosol Education Bureau and its western counterpart, the Aerosol Information Bureau. There was no group more aggressive in its defense of spray cans than the WAIB formed in mid-1975, to "keep the lid on bad publicity about aerosols in the eleven western states." The WAIB hired a Los Angeles PR firm, the Walter Leftwich Organization, which, over the next six months, inundated the media with press releases, brochures, reprints of articles, "situation papers," and press kits. According to Leftwich, the company had a "regular schedule of releases every ten days to two weeks."

Material was also sent to the science departments of five hundred high schools and the state colleges in the West. Leftwich believed that the impetus for concern about "so-called ecological problems" came largely from these students and "presenting our side of the story to them can do us a world of good."

The main thrust of the campaign was to ensure that stories "obnoxious to the industry" did not appear in the media and, by February of 1976, Leftwich seemed happy with the progress toward this end, saying that he was able to get promises from several California media organizations that they would not run stories without getting industry comment.

"It all worked and worked well," noted Leftwich. "For a period of some months all news about aerosols . . . was all but dried up. Days would go by without so much as a line about aerosols, and that is exactly what the game plan had been."

Some of the fluorocarbon producers did not leave all of the PR work to the industry associations. Kaiser Aluminum, for example, put out a slick twenty-page booklet called *At Issue: Fluorocarbons*. But the most intensive individual campaign was mounted, not surprisingly, by Du Pont, the world's largest single producer of fluorocarbons. The Du Pont program, according to one of the company's publications, involved "public-service advertisements, background meetings with key environmental, science, and editorial writers across the country, and monitoring of fast-breaking news." It was, of course, not long before several Du Pont scientists found themselves spending a large proportion of their time on the familiar treadmill of press conferences, congressional hearings, and speaking tours. Early in the game, Du Pont adopted a strategy of issuing "clarifications" to the media. As one of their own publications put it, in a rather superior tone, ". . . each new study or pronouncement is seized on as news and rushed into print. Each news story is treated as a victory or loss for the fluorocarbon/ozone theory. This adversary manner of discussion has lead to considerable public misunderstanding and fear. To avoid this, Du Pont is monitoring new developments . . . and issuing clarifications to the press, radio, and television where necessary."

Du Pont moved with breathtaking speed; the clarifications invariably followed hot on the heels of newly announced research results (particularly those damaging to the aerosol cause) or anti-aerosol pronouncements by public figures. At a meeting in Utah in September 1976, Du Pont carried this art form to a new high by issuing a rebuttal to a speech before the speech was delivered.[7]

The "clarifications" often adopted a scolding tone and frequently

[7] The speech, given by Russell Peterson, then chairman of the President's Council on Environmental Quality, called on industry to voluntarily phase out fluorocarbons. The night before his speech, a story in the Salt Lake City *Tribune* suggested that Peterson might say this. Early the next morning, as reporters and scientists entered the conference room to hear the speech, Du Pont's representatives were lying in wait with a press release protesting Peterson's position.

purported to tell the reader what a new piece of research "really" meant. As far as industry was concerned, it invariably meant that the Rowland/Molina theory was on shakier ground than ever before. In what respect this method of operation avoided an "adversary manner of discussion" was never quite clear.

In mid and late 1975, Du Pont took out full- and double-page ads in many large newspapers and magazines (described as "a series of discussions by Du Pont to offer a perspective on an important subject"). The headline on the first ad read: "The ozone layer vs. the aerosol industry. Du Pont wants to see them both survive." It was signed by Board chairman Irving Shapiro. The headline on the second read: "You want the ozone question answered one way or the other. So does Du Pont."

The scientific content of these ads will be discussed later, but it is worth pointing out here that the second of the ads contained a rather blatant example of a favored industry technique: selective quotation from the scientific literature. The ad quotes a paper by the Harvard scientists saying that *if* additional removal processes for fluorocarbons[8] or chlorine could be identified, the atmospheric and biological impacts of fluorocarbons "would be reduced accordingly" (emphasis added).

The point was a weak one anyway—the Harvard scientists had not said, or even implied, that such removal processes *did* exist. But the fact that Du Pont was really grasping at straws becomes particularly apparent upon careful examination of the Harvard paper itself. The quote used in the Du Pont ad comes from the last *footnote* of the paper in which the researchers caution that studies on the fluorocarbon problem were still at a preliminary stage. The Du Pont ad did not, however, quote another section of the same footnote, which said that these studies "develop a case . . . for serious concern. . . ." Nor did it quote the main conclusion in the text of the paper: "The results . . . raise serious questions which must be addressed as a matter of urgent priority by those responsible for public health and environmental policy. On the basis of best current models for the stratosphere and contemporary knowledge of Freon chemistry, there are reasons to believe that the present con-

8 The Harvard paper actually used the Du Pont trade name Freon, but it was carefully excised in the Du Pont ad.

sumption levels of Freon may pose serious problems for atmospheric ozone."

It should also be noted that the Harvard paper had been written in the fall of 1974 and, at that time, research on the problem was indeed very preliminary in nature. But by the time the paper was quoted in the Du Pont ad—nearly a year later—considerable additional effort had been expended by many scientists, including those in the industry. All attempts to convincingly identify a removal process for fluorocarbons or atomic chlorine (other than destruction of stratospheric ozone) had failed.

Throughout the fall of 1975 and the spring of 1976, Du Pont was churning out "clarifications" at a rate of at least two a month. Late in 1975, *Rolling Stone* magazine decided to try and find out what all this was costing Du Pont. "I don't have that figure," *Rolling Stone* quoted one Du Pont official as saying, "and I don't think we'll tell you anyway. People not working in this area could very easily misinterpret a dollar amount." *Rolling Stone* said that Du Pont has an annual corporate advertising budget approaching $30 million.

Du Pont refuses to release figures for the amount of money they've spent on the PR effort to save fluorocarbons. The reason, says Du Pont public-relations spokesman John Roberts, is that such figures would be "misleading" to someone who makes $10,000 or $15,000 a year. But $100,000 or $1 million "or whatever the number might be" is a "very small amount to spend" in defending a business that may be worth tens of millions or hundreds of millions of dollars, he said.

It is therefore difficult to get a figure for the total cost of the elaborate campaign Du Pont mounted in mid and late 1975, when they took out full-page ads in major newspapers and magazines across the country. However, based on the advertising rates at the time of the campaign, it would appear that Du Pont paid on the order of $25,000 for just three of the ads.

A spokesman for the Aerosol Education Bureau declined to give figures for the AEB's budget. On the other hand, a representative of COAS estimated that COAS's annual budget was on the order of $350,000 in 1975 and 1976. (This did not include the time and travel expenses of many industry speakers whose COAS activities were subsidized by their own companies.)

In the spring of 1976, Jack Dickinson, vice president of consumer affairs for Gillette, who had been the chairman of COAS, estimated that the whole industry had spent some $2 million in defense of aerosols. (This included money spent on toxicological research unrelated to the ozone issue.)

Janet Lowenthal wrote in *The Progressive* magazine that ". . . the very size and strategic placement of the Du Pont advertisement on the last page of the first section of the New York *Times* . . . reflect the ease with which a major U.S. corporation can pay the high publication cost of presenting its views to the American people, safe in assuming that no public-interest or conservation group has comparable financial resources to bring disinterested arguments before the public." (The industry, of course, argued that these groups didn't need advertising money because they got free coverage from a sympathetic press.)

The industry was, for the most part, unified in its attack on the fluorocarbon/ozone hypothesis. But solidarity broke down, particularly among the marketers, when it came to the very practical problem of advertising aerosol products. Consider the permutations. There were basically three types of products affected by the controversy: spray cans containing fluorocarbons, spray cans containing other propellants, and nonaerosol containers such as roll-ons or pump sprays. Different marketing companies carried varying combinations of these containers, and each company had to assess its own strategy and options in light of the commitment it had already made to fluorocarbons. Each had to ask itself whether the troubled propellants were worth fighting for.

A schism very quickly occurred between marketers who used fluorocarbons in their spray cans and those who didn't. Those who did became very cagey about admitting it, something they could get away with because they were not required to list the propellants on the can. The magazine *New Times* once wanted to run a picture featuring several different products and decided to call the marketers to find out if their cans contained fluorocarbons. One company told them the information was privileged. Another said the only man who could tell them that was away for several weeks.

Two referred the question to their lawyers. Only one company admitted that their product used fluorocarbons. Finally, *New Times* took all the cans to a lab for chemical analysis. Two were eliminated. The rest "passed with flying fluorocarbons."

Marketers whose cans did not contain fluorocarbons were particularly irritated when their products were used to illustrate newspaper or magazine stories about the ozone controversy, and they began a vigorous campaign to dissociate themselves from the fluorocarbon battle. They were worried not only about sales but also about the possibility of being drawn into a larger skirmish over spray cans per se. Theoretically, the spray can itself was not on trial here, but in fact, the cans were already under attack from consumer and environmental groups for other reasons, and the ozone problem simply gave them added momentum.

A manufacturer of valves for the food industry who was unaffected by the ozone problem asked in *Aerosol Age* whether it wouldn't be "better to narrow the controversy to the fluorocarbons. In that way, the taint would be taken off the entire industry." But some marketers did not bother to ask. Industry solidarity went out the window when several marketers aggressively started advertising the fact that their cans did not contain fluorocarbons. One company that sold paint in aerosol cans spent $40,000 printing several million copies of a "disavowal" kit, which included a placard for counter display saying that the aerosol paint products did not contain fluorocarbons "presently under investigation as possibly affecting the ozone."

The attempt to avoid guilt by association sometimes reached ridiculous lengths. One manufacturer of air freshener and insecticide sprays suggested that lumping spray cans that did not contain fluorocarbons in with those that did implied the kind of thinking that led the Nazis to execute an entire village "because they thought one or two people might be guilty."

The first major break in industry's ranks came in mid-1975, with the announcement by Johnson Wax that it would henceforth use no fluorocarbons in its spray cans in the United States. Samuel Johnson, president of the company, took full-page ads in many newspapers and magazines to announce that Johnson cans would carry the label: "Use with confidence. Contains no Freon or other fluorocarbons claimed to harm the ozone layer." (The word "Freon" was later dropped in response to a protest from Du Pont.) In a folksy "open letter to consumers," Sam Johnson (as he signed himself) said that he was taking this action "in the interest of our customers and the public in general during a period of uncertainty and scientific inquiry." He was careful not to express acceptance of the fluorocarbon/ozone theory: "Our own company scientists confirm that as a scientific hypothesis, it may be possible, but conclusive evidence is not available one way or another at this time." The message was clear, however: Not all spray cans should be tarred with the same brush.

Johnson's move was regarded by some in industry as a cynical attempt to exploit the ozone issue to gain a marketing advantage, and it caused a good deal of rancor. It came at a critical time, just when others felt they were getting control of their defense, and it caught many of them by surprise. It was clearly regarded as a hit below the belt, particularly since Johnson had been converting from fluorocarbons for some years and, at the time of the announcement, fluorocarbons made up less than 5 per cent of their total use of propellants. As Donald Davis noted in *Drug and Cosmetic Industry,* Johnson's action was understandable since sales of its products "were apparently suffering mightily because consumers couldn't tell one propellant from another. However, there was some criticism of the fact that S. C. Johnson used no fluorocarbons to speak of anyway and claiming to 'discontinue' a practice the claimee hasn't been following didn't seem like fair play to some competitors." Others in the industry were concerned that consumers might feel that if Johnson thought there was something wrong with fluorocarbons, it must be true.

The most blatant challenge to Johnson came from a competitor, Scott's Liquid Gold, Inc., a manufacturer of furniture polish. In an ad titled "Who's pulling the wool over whose eyes?," Scott's Liquid Gold pointed out that most of Johnson's products used hydrocarbon propellants "which are also potential contributors to illness and pollution." Scott's Liquid Gold, on the other hand, used carbon dioxide—"the environmentally safe propellant . . . This is the same gas that is used in *soda pop*." (Emphasis theirs.) The ad concluded by urging consumers to support companies that "show genuine concern for consumers and the environment in preference to those who make token adjustments."

Some marketers clearly decided that it wasn't enough to remove the fluorocarbons from their spray cans—that the spray cans themselves could not escape the taint of the ozone controversy. Many began converting to alternate packaging, such as pump sprays, roll-ons, or sticks. (One wit wondered if the police department would soon start using roll-on Mace.) By the end of 1974, one trade magazine was reporting a tremendous switch to pump sprays, which previously made up only 5 per cent of the market. They predicted that the public might abandon the aerosol spray within a year or two.

The marketers were clearly hedging their bets. Maybe there was nothing to the ozone problem; but, then again, maybe there was. The company scientists were fighting the good fight, but from the marketing point of view it hardly mattered who was right scientifically if the public turned against spray cans, for whatever reason. Maybe it was safer just to avoid the problem completely. The situation was perhaps best summed up by Ira Haskell of West Cabot Cosmetics, Inc. Announcing that his company would not package a new product in aerosol form, he said: "We put out good products. So why take a chance?"

The industry naturally tried to keep as much of the internal struggle behind closed doors as possible, but this was impossible to do when the switch to alternative packaging became the dominant theme in the marketers' advertising. By mid-1975 there was a proliferation of ads extolling the virtues of the nonaerosol. "No one can deny the blatant fascination for nonaerosol dispensers . . . by aerosol marketers," Donald Davis noted, as though speaking of some strange and shameful aberration.

If consumers recognized that this abrupt switch was due to the ozone controversy, they certainly did not discover this fact from the early ads themselves. The word "ozone" was never mentioned, and few of these ads made even the vaguest references to environmental concerns. Instead, there were euphemisms and coy circumlocutions like "the ideal antiperspirant for today" and "tomorrow's solution to problem perspiration today." To the *cognoscenti,* these ads were amusing; to those who did not know what was going on, they must have seemed rather confusing. You can see the dilemma the advertising people were facing. They didn't want to talk about the ozone problem in so many words—that would just spread the tale. Worse, it might give people the impression that the aerosol companies actually believed the ozone/fluorocarbon theory, and that certainly wouldn't do. Besides, there was always the possibility that the whole nightmare might just go away.

But it did not go away, and eventually the ads were forced to recognize the environmental issue. Though explicit references to the ozone problem were rare, some marketers featured "man in the street" scenarios in which people said they'd switched to pump sprays to help preserve the environment.

What dominated the ads, however, was not the ozone issue, but economics. The simple fact was that nonaerosols were much more economical than aerosols. Of course, this was just as true *before* the ozone debate started, but somehow the subject never seemed to come up in those pre-ozone ads.

A few marketers who sold both aerosol and nonaerosol products tried to advertise both at the same time and found themselves in a mildly schizophrenic situation. One ad, for example, went on at some length about the advantages of the nonaerosol. At the end, it featured animated aerosol and nonaerosol packages proclaiming, "We're both stars." This hardly followed, since the aerosol really hadn't had much to say for itself. But logic never was a strong suit in aerosol advertising.

Of course, the marketers who had abandoned the aerosol altogether had no such difficulties. In fact, they made mileage out of knocking spray cans; they would point out, for example, that the fluorocarbon propellants made up a large portion of the contents of an aerosol and that the roll-on or pump spray would last as long as two or three aerosols. Perhaps the most obnoxious example was an ad for a stick deodorant that carried the unsubtle and off-putting catch line: "Get off the can. Get on the stick." Another marketer took the *macho* approach. This ad featured a debonair European male, with a blonde clinging to him, telling his listeners how much he liked "The Pump."

These tactics naturally made many in the aerosol industry very irritable; one aerosol filler was moved to launch an advertising campaign called "Don't Kick the Can." The pages of the trade magazines were filled with gripes about ads that needlessly promoted bad feelings within the industry, ads that subtly gave credence to the doomsday predictions, ads that were "selling out."

Robert Abplanalp of Precision Valve Corporation took out a series of full-page ads saying that the "solidarity of the industry is important for the survival of all of us. . . . A united industry is a strong industry." (Production of aerosol valves had been cut in half by March 1975, and Abplanalp temporarily closed his headquarters plant in Yonkers, New York.)

Defending the spray can had to be a difficult and ultimately unrewarding task. It was an easy target for environmentalists—a technological sitting duck in the Age of Environmental Chic. Even before the ozone controversy came along, it was becoming increasingly fashionable to disapprove of spray cans. The can, said one industry spokesman, was the victim of "antitechnology snobbery." Despite industry's claims to the contrary, spray cans were, with the possible exception of asthma remedies and some other medicines, largely wasteful luxuries. For most people, they were also an easy sacrifice. Giving up spray cans was a change in life-style that was not particularly hard to live with; it had the perhaps unique advantage of being a virtually painless exercise in environmental responsibility.

Of course, the same could not be said of refrigeration and air conditioning. Defending these uses of fluorocarbons was much easier, and industry was not above using a few scare tactics. They claimed, for example, that defense computers would grind to a halt for want of air conditioning. They warned that existing refrigerators would not work with different coolants, that it would be hard to convert equipment to new coolants, and that nonpoisonous replacements "are not even on the drawing boards." One industrial chemist was quoted in *Business Week* as saying that if a ban were declared on fluorocarbons immediately "we would have to go back to delivering blocks of ice to refrigerate food, and air conditioning would be nonexistent." The industry persistently tried to plant in the public mind the idea that any move to regulate fluorocarbons in spray cans would necessarily also be an assault against the use of fluorocarbons in refrigeration and air conditioning. Thus did they try to shelter the spray can behind the protective shield of the refrigerator's unarguable social utility.

It wouldn't work, of course. No one had ever suggested that fluorocarbon refrigerants be banned without a grace period or that the nation's refrigerators and air conditioners be peremptorily dismantled. It was precisely *because* refrigeration had many more benefits to weigh against the risks that its case was different from that of spray cans. In any event, the spray-can propellants represented about 75 per cent of the problem, and some scientists suggested that an immediate ban on the propellants would buy time to tackle the refrigeration problem.

Many in the aerosol community were resistant to this kind of plea-bargaining for refrigerators. It did not escape their notice that the victim in any such "deal" would be the spray can. They clearly wanted the fluorocarbon-related industries, in the words of the famous phrase, to hang together and had no patience with the people who believed that destroying aerosols would not endanger air conditioning and refrigeration.

There seemed to be no tenet in industry's philosophy more unshakable than the one that held that the American consumer wanted, needed—indeed, loved—the aerosol spray can. (However, some in the industry have admitted that industry's marketing and advertising activities had played a role in convincing people they wanted and needed spray cans.)

Aerosol-industry spokesmen argued that the consumer would continue to pay extra to save a few moments despite being told that he or she was being overcharged for the aerosol product. In fact, there were those in the industry who were firmly convinced that the public would continue to buy spray cans in the face of all warnings about their harmful effects. "Despite the fact that the scientists and press may say that a product is harmful; despite the fact that they may even prove it is harmful—you still have to consider the fact that people want it," said Salvatore Noto, president of Technair Packaging Laboratories, Inc. "Sooner or later, the public gets what it wants, even in spite of precipitous [sic] legislation," he added, pointing out that cigarette sales had not been affected by warnings about the hazards of smoking.

One trade magazine asked whether the fickle consumer could really cast aside spray cans "like yesterday's love." The answer was, of course they could, if for no other reason than that the industry, after having spent so much money and effort talking people *into* using aerosols, was now spending a good part of its advertising dollar talking people *out* of it. Spray cans, while certainly ubiquitous, were far from addictive. As word of the ozone problem began to spread, it became increasingly clear that the American consumer did not need the cans all *that* much. One retailer reported that his customers were asking for "anything but aerosols." A marketer was quoted in *Packaging Digest* as saying that every aerosol product could be put into alternative packages with only minor problems of consumer acceptability.

At the end of 1974, production of aerosols was down by nearly 7 per cent from the previous year; the personal-products category was off by nearly 11 per cent. By the first half of 1975, sales of fluorocarbons had dropped nearly 20 per cent and production of aerosols was down 26 per cent. By mid-1975, the trade magazines were reporting "impressive gains for nonaerosols" as the alternative packages began to seriously erode the overwhelming dominance of the aerosol in the marketplace.

At the end of 1975, production of fluorocarbons was off 15 per cent and the production of aerosols was down 13.5 per cent, with personal products dropping nearly 20 per cent.[9]

Although they are indicative, these statistics cannot be taken as direct evidence for the impact of the ozone controversy. Industry spokesmen, in fact, were inclined to dismiss the ozone problem as "overdone" and attribute the aerosol's dwindling fortunes to other factors. One suggested that people were wearing new hairstyles that required less hair spray. Another suggested that it was hard to find anything new to put into spray cans and that it might actually be unrealistic to hope for limitless growth.

But the prevalent theory was that the dismal statistics for 1975

[9] By the end of 1975, European countries were also reporting declines in aerosol production ranging from 7.8 per cent to 24.6 per cent.

reflected the deteriorating state of the economy more than any consumer awareness of the ozone issue, and several people in the industry predicted a resurgence of aerosol sales with the end of the recession.

The trend of 1975 was almost certainly influenced by economic factors—and, no doubt, the industry's ads pointing out that a nonaerosol container was worth two or three aerosol cans soon began to take their toll—but is it possible to determine how big a role the ozone controversy actually did play in influencing aerosol sales?

Some clues can be found in informal surveys published in newspapers and magazines. Here opinion ran the gamut. Some people still believed that the atmosphere was an infinite resource; a Philadelphia dentist remarked that he did not see how "a little squirt like that will make any difference in a big environment like ours." Other people were tired of these predictions of doom and disaster. "I'm beginning to wonder if so many things are harmful, why we just don't all stay in bed. But then we'd all get bedsores."

But there was increasing evidence that the ozone problem genuinely worried many people and that they were giving some thought to what they should do about it. One consumer, for example, asked Schiff it it was all right to use her hair spray if she kept the bathroom door closed to prevent the fluorocarbons from escaping. (It won't work—unless, of course, you have a hermetically sealed bathroom.)

The aerosol companies also began to hear from the public. Sol Ganz, president of Bronze Powder, Inc., was a bit shaken by a letter he received from a young Maryland girl. She said that she and her friends didn't want to die of skin cancer twenty years hence. Moreover her horse's foal, about to be born, would also be alive then "and I'll probably have to shoot her." She said Ganz could stop this by switching to pump sprays—or didn't he care?

Horrified, Ganz wrote a lengthy letter back. He pointed out that some 50 per cent of spray cans (including the ones his company produced) did not contain fluorocarbons. He argued that the ozone/fluorocarbon theory was still only a theory, and that Congress was not yet convinced there was a problem. He concluded by saying that he did not believe the girl's health and that of her foal were in danger.

It is, of course, always difficult to tell from isolated examples such as these whether or not a trend is developing. Some in industry firmly believed that there was no real anti-aerosol trend, that the sentiment was and would remain a very limited phenomenon. One industry official, quoted in *Food and Drug Packaging* magazine, said: "I think only a small percentage of consumers have stopped buying aerosols. These are the bright kids in their twenties who are always on some bandwagon or other."

However, at least two formal public-opinion surveys were conducted on the fluorocarbon issue. One was done by a market-research team from Du Pont, which had been studying the impact of bad publicity on the public's buying habits since 1972.

The 1975 statistics for this study were collected after the ozone controversy had become big news. More than half the respondents in March 1975 indicated they were aware of the controversy (compared with a quarter of the respondents questioned just four months earlier, in November 1974). Twenty-nine per cent of the respondents in 1975 said they were using aerosols less (compared with 16 per cent in 1974). Sixty per cent said they were using about the same number of aerosols.

The study revealed that, of the 29 per cent who said they were using aerosols less, nearly half cited "danger to the environment" as a reason, second only to "danger to health." This group showed a strong awareness of the ozone issue, the study reported.

Another survey was conducted in November 1976 by De Vries and Associates, Inc., for the Consumer Product Safety Commission, an agency of the federal government. Some 1,800 people aged 18 or older were interviewed by telephone. The respondents in this survey, informed of the theory that fluorocarbons would damage the ozone layer, opted 60 per cent in favor of taking spray cans off the market. Nearly 30 per cent favored labeling fluorocarbon-containing cans. Two thirds of those questioned said they would not be bothered at all if fluorocarbon-containing spray cans were taken off the market and replaced with alternative products. Nearly 30 per cent said they would be slightly inconvenienced, and only 2.8 per cent felt they would be greatly inconvenienced.

Respondents were also asked if labels on fluorocarbon-containing spray cans would influence their buying habits. Seventy-two per cent said yes. About 20 per cent said no, but more than a quarter of these people said this was because they had already stopped using aerosols. (The rest said it wouldn't affect their buying habits because the hazard hadn't been proved or because "I don't care . . . it doesn't matter to me.")

The survey showed that, by late 1976, a large proportion of the American public (73.5 per cent) had heard something about the spray-can issue. (The media—primarily newspapers—proved overwhelmingly to be the sources of their information.) About a quarter had not heard anything. Of those who had heard, somewhat less than half used the word "ozone" in their answer. Others referred more vaguely to the fact that spray cans might be harming the environment or the atmosphere. Some respondents were able to specify that fluorocarbon propellants were at issue, and others also mentioned ultraviolet radiation and cancer.[10]

The survey revealed that half of American consumers say they had not stopped using products because of what they had heard or read, while nearly 45 per cent said they had. Of those who said they had stopped using certain products, half named aerosols.

However, the study indicated that more than half (55.2 per cent) of Americans had decreased their use of aerosols, and nearly 40 per cent of those did so because of the possible damage to the ozone layer and increases in ultraviolet radiation and cancer.

In other words, about one fifth to one quarter of the American people had actually done something about it.

As might be expected, however, there seemed to be a large number of people who were waiting for the government to act. The feeling was summed up by one consumer quoted in a New York *Times* story: "Until they take aerosols off the shelves, they seem to be saying it's all right to use them."

---

[10] The public could, however, acquire some rather startling misconceptions about the ozone controversy. In one study, some people reportedly said they believed their underarm deodorant spray contained the harmful chemical ozone which could produce skin cancer.

"I think they should do something so the aerosols won't hurt me," said another.

The aerosol industry tried to capitalize on this feeling in its advertising. As an ad for one aerosol filler put it: "The fact that you can purchase aerosols using fluorocarbon propellants in almost every store in the U.S.A. is the best proof that to date no one has proven that these propellants have harmed people or our environment."

This, of course, was what the political battle was all about. The industry launched a vigorous campaign to stave off government regulation, or at least to postpone it as long as possible. After all, winning the public-relations battle with the consumers would count for nothing if spray cans were legislated out of existence.

CHAPTER EIGHT

To Ban or Not to Ban

In view of . . . the very substantial benefits attributable to the use of fluorocarbons, as well as the severe economic consequences of any regulation, we at Du Pont feel that legislation aimed at such regulation is unwarranted at this time.

> —Ray McCarthy, Du Pont, testifying before a
> congressional subcommittee, December 12, 1974.

Du Pont is just stonewalling.

> —Sherry Rowland, in *The Nation* magazine,
> June 28, 1975, page 775.

December 11, 1974, marked the premiere performance of the Incredible Stratospheric Traveling Road Show and Debating Society. The society was composed of university researchers, aerosol industry scientists and spokesmen, and representatives of government research organizations and regulatory agencies. They not only put on a long-running performance in Washington—the Broadway of the governmental hearings circuit—but also, when state legislatures beckoned, they took the show on the road. On December 11, in Washington, the House Subcommittee on Public Health and the Environment opened its investigation into the fluorocarbon issue; this was to be only the first of dozens of similar hearings that would be held all over the country in the next two years.

Politically, the ozone problem was bad news in every way. It was global in nature (described, in fact, as possibly the first truly global environmental problem). The consequences were potentially serious, but neither immediate nor obvious. The scientific uncertainties were large—and so were the potential economic penalties. And, to complicate matters further, there were competing pressures from all sides—from scientists, regulatory agencies, and industry.

At the most fundamental level, the political issue was reduced to only one question—whether, and when, to restrict or ban the use of fluorocarbons. This issue was debated primarily in testimony before government committees, but also on the public-speaking circuit and in the media.

If any single theme could be said to characterize industry's pitch, it was that the Rowland/Molina theory was just that—a theory. It had not been "proved." The industry sought to persuade Congress and the regulatory agencies that it was safe to wait two or three years before taking any action so that more data could be gathered.[1] Industry spokesmen even strongly implied that no political action should be taken until the industry's own research program was completed at a cost of some $3 million to $5 million two or three years hence.

Sherry Rowland did not agree. He was not at all impressed with industry's $3 million to $5 million, nor with the case they were making for a two-to-three-year delay in regulations. He pointed out that the U. S. Government had spent perhaps as much as $100 million during a three-year CIAP program to gain considerable understanding of the stratosphere, and he did not see why everyone had to wait with bated breath while the industry's scientists, who had never before done anything on the ozone question, pondered the problem. Rowland came out early in favor of a ban on fluorocarbon-containing spray cans—a position from which he never wavered—and this firmly established him in the anti-industry camp. Ralph Cicerone, also vocal on the subject, soon joined him, and it

[1] Industry spokesmen persistently warned the government against taking what they almost invariably called "precipitous action." They were fond of pointing out cases of too-hasty regulation, and a favorite example involved the flap over mercury levels in tuna. They repeatedly emphasized that the same mercury levels had been found in tuna nearly a century old, but apparently never realized they were talking about Sherry Rowland's work.

was not long before the two of them were jointly and severally branded as the aerosol industry's version of Ralph Nader.[2]

As early as the beginning of 1975, Rowland was arguing that there was sufficient scientific understanding of the stratosphere that "it's beginning to seem unlikely that there is a major flaw" in the calculations concerning the impact of fluorocarbons. The *Nature* paper had been out for nine months, and it was circulated in scientific circles for five months before that, he pointed out. Several groups of scientists had done independent calculations and all confirmed the indications of serious environmental damage. "The search for the major mistake has been remarkably unsuccessful. How long do you go on pursuing this will-o'-the-wisp of a major mistake?" In light of the consequences, he said, "I think we've got to start behaving as though the theory were right."

Industry, of course, considered this to be lynch-party tactics of the worst sort. And so the battle was joined.

On December 11, 1974, Paul Rogers of Florida, chairman of the House Subcommitee on Public Health and the Environment, started the ball rolling by calling hearings on the ozone question. Rogers and another congressman, Marvin Esch of Michigan, had jointly introduced a bill to amend the Clean Air Act, calling for the National Academy of Sciences to do a major study of the fluorocarbon problem and authorizing the Environmental Protection Agency (EPA) to ban the chemicals if indeed they proved to be a threat.[3]

This bill and the hearings associated with them touched on one of the urgent problems facing the political system in its attempts to deal with the fluorocarbon/ozone issue. It was all very well for

[2] In the case of Rowland, one dissenter was Donald Davis, editor of *Drug and Cosmetic Industry*, who wrote that he had seen Rowland reject opportunities to condemn all aerosols or to use inflammatory language just to give reporters better copy. "Rowland," he said, "apparently has no ambitions toward becoming a Ralph Nader." Davis wasn't so sure about Cicerone, however.

[3] Similar provisions were later incorporated into a bill introduced into Congress. It was debated for about two years, during which time enactment seemed the probable outcome. However, this amendment and others were unexpectedly defeated when the bill was filibustered off the floor in October 1976. Amendments providing for a major study of the stratosphere were finally passed in the summer of 1977.

Congress to ponder whether or not to regulate fluorocarbons, but in fact it was not at all clear who should—or could—actually do the job and how. The regulatory situation was, to put it bluntly, a morass—something that had become immediately obvious at the outset of the fluorocarbon controversy when the Natural Resources Defense Council (NRDC) made the first move against the spray cans.

The NRDC, a legal organization that initiates actions relating to environmental protection, petitioned the federal Consumer Product Safety Commission to ban fluorocarbon-containing spray cans, a pitch based largely on the skin cancer effects of ozone depletion.

However, despite its name, the CPSC was perhaps not really the appropriate body to receive this petition. Regulatory control over fluorocarbons was, at that point, extremely fragmented, and the CPSC did not have jurisdiction over most of the spray cans that contained fluorocarbons. Hair sprays and deodorants, for example, were the province of the Food and Drug Administration. The FDA, which controlled all drug, food, and cosmetic products, had jurisdiction over about 40 per cent of all fluorocarbons produced in the United States and about 80 per cent of the fluorocarbons used in spray cans. The Consumer Product Safety Commission did have jurisdiction over propellants used in household-cleaning products, but many of these probably did not contain fluorocarbons. Moreover, the CPSC had authority over the fluorocarbons in refrigerators used in homes and schools, but not over those used in car air conditioners or in industrial applications. (In fact, no one had regulatory control over the latter.)

To add to the complications, the CPSC had been directed to defer to the Environmental Protection Agency whenever the EPA could take action under the Clean Air Act (although the EPA's authority to do so in the case of the fluorocarbon threat to the ozone layer was, at that time, iffy at best).

At the Rogers hearing, industry's defense was handled largely by Du Pont's Ray McCarthy. He told the congressmen that the industry directly dependent on fluorocarbons would contribute more than $8 billion to the U.S. gross national product that year and employ some 200,000 Americans (figures, as we have seen, that are subject to some interpretation). McCarthy acknowledged that a ban on fluorocarbons might not affect all 200,000 jobs, but "it would cause tremendous dislocation, particularly in the short run."

Arguing that the fluorocarbon/ozone theory was "purely speculative," McCarthy stated that immediate regulations to restrict the chemicals were unwarranted. Instead, he urged that an experimental program be set up to discover the truth, and concluded his testimony with the following pledge: "If creditable scientific data developed in this experimental program show that any chlorofluorocarbons cannot be used without a threat to health, Du Pont will stop production of these compounds."

The hearings ended on December 12. The political system being the efficient machine it is, the lame-duck session of Congress expired eight days later without the bill being reported out of the subcommittee, guaranteeing that the entire process would have to be repeated in 1977.

On January 29, 1975, the Senate Committee on Aeronautical and Space Sciences got into the act. The first speaker was Tom Donahue of the University of Michigan. He emphasized that it had taken hundreds of millions of years for life on earth, its atmosphere, and its protective ozone shield to develop as a single interdependent system. Large changes in that system taking place over a shorter time scale would have a "disastrous effect," he said. "We appear to be on the verge of a period of great peril to life on this globe. . . ."[4] Donahue emphasized that there were uncertainties in the calculations but said that scientists knew of "no obvious deficiencies" in the calculations. The calculations had been successful in explaining the properties of the normal atmosphere and were "based on well-studied and thoroughly understood chemical reactions, for the most part. We may be as much as a factor or two wrong in our time scales or our magnitudes but it is highly unlikely that we are completely wrong in projecting qualitatively the nature of those effects."

[4] Donahue said that an ozone reduction of 40 per cent (then the calculated impact in the Northern Hemisphere of a fleet of five hundred Boeing-type SSTs) "would probably drive life on the globe back toward a state it had several hundred million years ago. . . ." As we shall see, this later caused him some difficulties.

Mike McElroy followed Donahue to the witness stand.[5] McElroy argued that there was an urgent need to build up the country's scientific expertise in atmospheric chemistry and that several tens of millions of dollars ought to be devoted to the task.

Even among nonindustry scientists (excepting Sherry Rowland), a consensus seemed to be emerging that it would not do irreparable harm to delay a political decision for a year or so to allow further scientific studies to be carried out. However, Lester Machta of the National Oceanic and Atmospheric Administration warned that there could be no assurance that a three-year national research program would produce unambiguous results. Nevertheless, he took the position that the seriousness of the problem was such that, unless strong evidence was produced to refute the theory, regulations would be necessary at the end of that time anyway.

In February 1975, the Rogers/Esch bill that had died in committee in the previous session was reintroduced. A second bill was also introduced in the House, by Congressman Les Aspin of Wisconsin. This bill had some unique features. It not only proposed a ban on fluorocarbon propellants, but also called for limits to be placed on the total amount of fluorocarbons that could be made or imported into the United States. Licenses for these amounts would be sold at auction. The bill also called for the EPA, with the advice of the National Academy and NASA, to establish limits on fluorocarbon releases to the atmosphere based on the EPA's judgment of an acceptable level of ozone reduction. As a temporary limit, the bill set the release rate at 10 per cent of the then current rates. This bill did not pass.

Interestingly, the question of just how much ozone reduction would be "acceptable" has never been resolved. No political or scientific body anywhere in the world has yet made this judgment. Most scientists, however, see it as essentially a political decision—

[5] The relationship between McElroy and Donahue had been steadily deteriorating since Donahue had gone to Michigan. McElroy opened his testimony by saying that although he had been asked to co-ordinate his statement with Donahue, "communication between the state of Michigan and the commonwealth of Massachusetts appears to occur with great difficulty." McElroy said Donahue had promised him a copy of his (Donahue's) testimony, but "neglected to point out that he was only going to deliver the copy to me this morning."

somewhat akin, as one put it, to the congressional decision to impose a fifty-five-mile-per-hour speed limit.

As the hearings wore on, the pages of the aerosol trade press were increasingly filled with synopses of the battle, under such headlines as: "Ordeal over Ozone Depletion Continues." By and large, the trade press preferred to see the hearings not as an expression of concern for the environment, but as a kind of technological witch-hunt conducted by politicians, regulators, and scientists seeking political gain or personal publicity. *Aerosol Age* was not above a few catty comments. After a brief discussion of the skin-cancer issue, in which it was noted that only a small proportion of the world's population is susceptible, *Aerosol Age* commented that one public official at a hearing was fair-skinned and red-haired. This is, of course, precisely the type of person most susceptible to skin cancer. *Aerosol Age* seemed to be implying that such people would automatically favor an immediate ban, but the writer must have failed to notice that Mike McElroy who was regarded as rational and effective is also fair-skinned and red-haired.

With all the hearings—and the legislative bustle they entailed—going on, it was not long before the executive branch jumped on the bandwagon. There were, in fact, no fewer than eight government departments and agencies doing stratospheric research related to the ozone problem. In fiscal 1975, they were spending more than $14 million on this research, although little of this was really "new" money allocated solely because of the fluorocarbon problem.

It can be seen from this that jurisdiction over stratospheric research was as fragmented as was the regulatory power over fluorocarbons. One of the early political questions that had to be faced in the fluorocarbon debate was how best to marshal the nation's scientific expertise to cope with a problem that cut across so many territorial boundaries and affected so many entrenched interests. As Dale Bumpers of Arkansas, chairman of the Senate Subcommittee on the Upper Atmosphere, put it during one congressional hearing: "I see so many turf fights take place around here about who is going to be the lead agency and who is going to have the responsibility. . . . It does not serve the best interests of the

country where zealots get involved about who is going to do what
. . . the normal inclination is to keep the empire intact."

It was NASA that made the strongest and ultimately the most
successful pitch for "lead agency" status. As we saw in Chapter Six,
the space agency decided, after getting over its fright about the
shuttle, to diversify into the stratospheric game in a big way. As
early as January 1975, Fletcher said that "we have rather dramatically changed our emphasis within NASA to focus on stratospheric
research."

The tenor of congressional testimony given by various NASA
officials at this time suggests that the agency may have been motivated less by a strategy of simply trying to increase its own budget
than by a desire to find a new *raison d'être* for NASA. While not
actively discouraging the idea of additional funding, NASA emphasized that leadership was more important. By taking the lead in
tackling a problem of urgent public concern, NASA was perhaps
trying to still the criticism of those who continued to grumble about
what they considered to be useless extravaganzas in space. And, at
least in part, this tactic of not asking for more money also seemed
calculated to protect the very expensive shuttle program. It could
be argued that if stratospheric research had such a high priority,
NASA should perhaps be transferring funds from the shuttle rather
than seeking new funds. Looking at NASA's proposed 1976 budget
for the stratosphere, one senator remarked: "When you consider
the size of NASA's budget for the space shuttle . . . this seems like
peanuts, does it not?"

"Yes, it seems like a small amount of money compared to the
other," said James Fletcher, then NASA administrator, "but on the
other hand, as I mentioned in previous testimony, the primary
problems are not going to be solved by the application of large sums
of money."

Whatever the motivation, a number of significant changes had
occurred in the space agency by early 1975. NASA moved quickly
to consolidate its own rather disjointed and far-flung stratospheric
research effort into a single program that was funded in fiscal 1975
and 1976 at a level of about $7 million annually. NASA also
acquired a new advisory committee, the Stratospheric Research Ad-

visory Committee, composed of experts from universities and other outside institutions.

In addition, the National Aeronautics and Space Act was amended by Congress to authorize NASA to conduct a program of research, technological development, and monitoring of the upper atmosphere "so as to provide for an understanding of and to maintain the chemical and physical integrity of the earth's upper atmosphere."

Where before, when the shuttle had seemed menaced, NASA had kept a low profile on the ozone question, it now began aggressively to advertise its special competence to manage the whole show. (NASA officials unashamedly capitalized on the fact that the agency had, after all, been studying the ozone problem for 2½ years, although they neglected to mention that for part of that time, the effort was conducted almost clandestinely. One is left to wonder what NASA would have done with the results of those studies if the shuttle problem had turned out badly.) NASA trotted out the arsenal of technology it possessed to tackle the ozone problem. It had balloons, aircraft, rockets, and satellites to make atmospheric measurements. It even had—you guessed it—the space shuttle. It had computer models of the atmosphere. It had more; it had the cosmic perspective of the earth as a planet. It had the special expertise that came from studying the atmospheres of Venus and Mars. Why, NASA was even studying the sun itself and its relationship to the earth. Only NASA could call half the solar system to its defense. There are few agencies that can match it for sheer razzle-dazzle.

In mid-1975, the Interdepartmental Committee for Atmospheric Science (ICAS)—a group representing a number of government departments and agencies—appointed NASA the lead agency for the purpose of developing and testing the instruments needed for stratospheric measurements. NASA was also put in charge of an ICAS subcommittee that was charged with developing an expanded stratospheric research program for the government.

Of course, the other departments and agencies did not vacate the field. They not only continued their stratospheric research programs, they were also careful to point out that NASA was not really running the whole show. For example, in congressional testimony,

Robert White, then head of the National Oceanic and Atmospheric Administration, noted that the complexity of the ozone problem called for the talents and expertise of many different groups and "I don't think there is any single agency . . . that possesses all those competences and talents." He argued for co-ordination rather than centralized control of the research programs. ". . . I am not so sure you get where you want to go by having a single agency doing all the management in such a complex problem."

There was, therefore, a certain inevitability about IMOS. The acronym is short for the (Ad Hoc Federal Interagency Task Force on the) Inadvertent Modification of the Stratosphere. We saw from the Kyoto incident that when the scientific groups got into jurisdictional battles over the stratosphere, they often disdainfully refused to fraternize with one another. The government agencies, on the other hand, opted for the favored Washington solution—they formed a committee. IMOS was wholly a creature of the ozone debate. It was cosponsored by the President's Council on Environmental Quality and the Federal Council on Environmental Quality and the Federal Council for Science and Technology. It represented the departments of Communication; Justice; Health, Education, and Welfare; Agriculture; Defense, and Transportation; the Environmental Protection Agency; the National Science Foundation; the Food and Drug Administration; NASA; the Energy Research and Development Administration; the Consumer Product Safety Commission; and the Interdepartmental Committee for Atmospheric Sciences. (You now had one interdepartmental committee represented on another interdepartmental committee.)

IMOS was charged with doing a study of the seriousness of the fluorocarbon/ozone problem. In addition to the scientific question, IMOS also analyzed the state of the government's various research programs and examined the adequacy of the federal authority to regulate fluorocarbons.

On February 27, 1975, it held a one-day public meeting to hear testimony from the scientific community and other interested parties. Spokesmen for the Natural Resources Defense Council, in particular, urged IMOS to resolve the jurisdictional disputes so that they would not delay regulatory action.

Both Sherry Rowland and Mike McElroy were in attendance at

the IMOS hearing, and they got into a protracted and increasingly acrimonious wrangle over several issues. There was, for example, a testy exchange concerning the accuracy of the calculations of ozone depletion. McElroy suggested that there were quite large uncertainties in the calculations. Rowland thought McElroy's estimates of these uncertainties much exaggerated. In retrospect, McElroy believes that they both went overboard. "It got to the point where I was making statements about the inaccuracies in calculations that were somewhat extreme and he was making statements about the precision of the calculations that were overly extreme."

But this was not the only source of irritation. What really sparked trouble was the issue of fluorocarbon regulation. Rowland was becoming increasingly hawkish about the need for an immediate ban. McElroy, on the other hand, felt strongly that a ban was premature, given the uncertainties in the calculations. At the hearing, he argued that "if we stop using chlorine compounds within five years we will not have done irreparable harm to the environment. That is not to say we must not get moving and take it seriously. . . ." Five years was the longest waiting period to be suggested by anyone up to that point. Even the industry's representatives had been talking about two to three years (although their two to three years was a floating figure; two years into the debate they were *still* talking about an additional two-to-three-year waiting period).

To many people it seemed as though McElroy was backpedaling on the regulation issue. He subsequently denied that he had changed his position, and it is certainly true that as early as the fall of 1974, he had come out explicitly against an immediate ban. But at the same time, he had been explaining in rather graphic terms the penalties involved in delaying regulations—these were indicated by his computer study—and perhaps in consequence of this, his message seemed contradictory to many people. One trade magazine noted that McElroy "one day seems to side with industry, the next with Rowland and Molina."

In any event, at the time of the IMOS hearing, many people got the impression that McElroy had changed his tune on regulation. Walter Sullivan in the New York *Times* wrote that McElroy had "altered his earlier estimates, now stating that a ban on aerosol sprays might be delayed a few years without disastrous results."

Not surprisingly, this sort of interpretation was also advanced by industry representatives. Frank Bower of Du Pont, for example, offered the view that "Dr. McElroy of Harvard said the more frightening calculations of his group had been refined and that a brief delay in a decision on the Freon issue seemed acceptable."

This in itself might not have raised the hackles of those who supported an immediate ban so much were it not coupled with the fact that industry officials had discussed with McElroy the possibility of setting up an independent scientific committee to advise industry on its stratospheric research program. McElroy had suggested the committee to industry officials over lunch one day when he was visiting Du Pont to give a talk. He said that such a committee would help them avoid spending money on poor research. Du Pont was enthusiastic about the idea and wanted to follow it up. But Ray McCarthy, head of Du Pont's Freon Products Division, realized, as he put it, that "there's no way if I tried to put something together that anybody's going to think it's anything but a put-up job." So Du Pont asked McElroy if he would head the committee and recommend other scientists to sit on it.

The idea received much discussion in industry circles. At an industry gathering held the night before the IMOS hearing, according to one participant who was there, one industry official commented that "he felt it was important for the [fluorocarbon] producers to get an academic spokesman of great stature to present their view and he felt they had McElroy lined up."

At the IMOS hearing the next day, Ray McCarthy mentioned the industry's interest in setting up an advisory committee and having McElroy organize it. This contributed to serious trouble between Rowland and McElroy after the IMOS meeting. In conversation with another scientist, Rowland criticized McElroy for undercutting his position. Rowland says he does not recall ever suggesting that Mike's stand on regulations might have something to do with the apparent Du Pont connection, but this was evidently the impression McElroy got after hearing about Rowland's remarks, and he was outraged. He pulled out of discussions with Du Pont immediately and threatened Rowland that Harvard University would institute a lawsuit if any more aspersions were cast. Industry officials kept after Mike for a while but, according to McCarthy,

McElroy started giving them the cold shoulder, for reasons that were not explained at the time. "I didn't know why, but Mike just plain lost interest in it. [His] interest dropped off very rapidly . . . it went to zero." McElroy says he decided he couldn't do it because of this "backbiting and rumor business" but "I probably should've done it."[6]

It was shortly after the IMOS meeting that Rowland's spirits hit their nadir. His relationship with McElroy went rapidly downhill, and the business of the threatened lawsuit depressed him. Moreoever, he felt that the big guns in stratospheric research—including NASA—were ranged against him.

McElroy was also having problems in the aftermath of the IMOS hearing. At the hearing, he had temporarily thrown things off the track by suggesting that bromine, a better ozone-eater than chlorine, could be used as a weapon if injected over enemy territory in sufficient quantities. He urged an international treaty to ban such warfare. The statement tended to distract attention from the main issue at hand, which was the fluorocarbons, and it caused some puzzlement among those in attendance.

It was not very surprising that McElroy made headlines.[7] The New York *Times*, for example, reported: "Ozone Depletion Seen as War Tool." The most sensational treatment was found in the *National Enquirer*, which is not famous for restraint in such matters. Its headline read: "Harvard Professor Warns of . . . the Doomsday Weapon. . . . It's Worse Than the Most Devastating Nuclear Explosion—and Available to All." McElroy was quoted in the arti-

[6] McElroy continued to be sensitive about connections with industry. When a graduate student, Nien Dak Sze, abruptly left Harvard and went to work for a private company that was later contracted to do computer modeling for Du Pont, McElroy vehemently disclaimed any prior knowledge of this arrangement or any connection with the work. By the time Dak Sze left, the relationship between the two men had deteriorated because of a disagreement over the importance of Dak Sze's contribution to the early fluorocarbon/ozone work at Harvard.

[7] McElroy says he did not know his remarks were going on the public record. He said he was carried off to sign some papers just as the IMOS meeting was getting under way and did not hear the chairman's warning that the media were present and that the proceedings were public. However, another who attended the hearing said McElroy apparently recognized the presence of Walter Sullivan, the New York *Times* science writer.

cle as saying: "A few kilograms of bromine is all that would be needed for a large, devastating effect. The delivery would be no problem. A small rocket, an aircraft, even a balloon would do. Any country in the world could handle it. And the terrifying thing is that right now, there's nothing to stop them." He is also quoted as saying that a bromine bomb "would be a doomsday weapon because it would cause equal harm to friend and foe alike." (This, in fact, is a point raised by those who feel such a weapon would never be used because of the compelling deterrent of self-interest.)

In an interview, McElroy stated that he has "absolutely no recollection" of talking to the *Enquirer* about bromine.

In any event, McElroy's introduction of the bromine issue seriously annoyed many of his scientific colleagues, who thought it scaremongering of the worst sort. "That was a disgrace," said one. Moreover, it was done at a delicate juncture in the ozone debate— at a time when many of the scientists involved in the fluorocarbon controversy were striving to establish their credibility and to assure the public, political leaders, and industry that they were not capriciously or needlessly crying doom.

This need for credibility was the subject of a strongly worded commentary in *Science* magazine. It was written by staff writer Allen Hammond, who was critical of the "alarmist statements" being made by scientists working on the ozone problem. "These statements are incautious to say the least, in view of the uncertainties still attached to the calculations and the assumptions on which they are based. Ironically, the eastern-university scientists who made them have also played major roles in discovering and documenting the vulnerability of the ozone layer to human activities—in establishing, in other words, that there is a real ozone problem. Whether alarmist statements can be attributed to what one scientist described as 'the smell of a Nobel Prize' or simply to poor judgment, they have not served to increase the credibility of a serious problem."

The fact that the article mentioned "scare statements" about bromine weapons, coupled with the phrase "eastern-university scientists," had the effect of singling out Mike McElroy, which is something Hammond had not intended. Hammond explains that the word "eastern" was added to his story without his knowledge after it had been handed in. Mike's comment about bromine had been one of the factors that prompted Hammond to write the story, but it was by no means the only one. Hammond felt that a number of the scientists in the field were "overdoing it" and his story was directed at all of them, not specifically at McElroy. "I felt they were risking their credibility on what I viewed as a genuinely serious problem."

In an early draft of the story, which he sent to various scientists for comment, Hammond named names, including McElroy's. This elicited a call from Tom Donahue. The article as it finally appeared included in its criticism a comment Donahue had made at a public hearing (the one about driving life on earth back hundreds of millions of years), but Donahue was arguing for Mike's sake, not his own. He felt protective about Mike; they had once been close colleagues and the very best of friends, although Donahue's move to Michigan had put a serious strain on the relationship. Donahue called Hammond at home on a weekend, catching him just as he was about to leave on a trip. He urged Hammond not to focus the article on McElroy, arguing that, despite the bromine incident, Mike deserved considerable credit for helping to alert the public to the ozone problem and that he was a person who, in the future, could contribute a great deal to its solution. Hammond had already decided to do this, and he told Donahue he would rewrite the article.

The statement about "the smell of a Nobel Prize" caused quite a stir within the scientific community. It was an apt description; the ozone game was tense and grimly competitive in the early part of 1975. "These people were working seven-day weeks and around the clock," said Hammond. "I'd seen that before . . . it was very clear what was going through some of these people's heads."

It was, as one scientist described it, the "Watson-Crick sort of thing."[8]

[8] The reference is to James Watson and Francis Crick, who won the Nobel Prize for their discovery of the structure of DNA, the genetic molecule. Prior to its discovery, scientists suspected that this was probably Nobel Prize material, and Watson and Crick, working at Cambridge University, were in a dead heat with the renowned American chemist Linus Pauling. In his famous and irreverent book, *The Double Helix*, Watson relates telling Pauling's son that he and Crick were racing Pauling for the Nobel Prize.

There was much speculation concerning McElroy's motivations in the bromine incident and other episodes. There were doubtless many reasons, but one theory that finds credence among those who know him is that he was having a hard time adjusting to being a Johnny-come-lately in the fluorocarbon/chlorine controversy. Consider that McElroy was a leading aeronomer and, because of his work on the atmosphere of Venus, an authority on chlorine chemistry. Yet he had essentially missed the hottest issue ever to come along in ozone/chlorine chemistry. Of course, so did all the other expert aeronomers—in this case, as with the SST, the problem was first identified by "outsiders"—but McElroy seemed to take it harder than most. He is not a *"c'est la vie"* kind of person. He clearly did not like being a bridesmaid, and his colleagues sensed, in many of his actions during this period, a sometimes rather frenzied attempt to recoup.

Four months after its public hearing, in June 1975, IMOS issued its report. IMOS concluded that the fluorocarbon/ozone problem presented "legitimate cause for concern." Unless new scientific evidence was found to remove that concern, restrictions on fluorocarbons 11 and 12 might well be necessary.

The report pointed out that the National Academy of Sciences was currently engaged in an in-depth study of the fluorocarbons and would report in less than a year. If this study "confirms the current task-force assessment, it is recommended that the federal regulatory agencies initiate rule-making procedures for implementing regulations to restrict fluorocarbon uses." It was necessary to start these procedures well in advance of the date that the regulations were to take effect because the process could take as long as two or three years. The report suggested that regulations could reasonably be effective by January 1978—enough time for additional scientific studies and for both the industry and the public to adjust to the changes. The task force also recommended that fluorocarbon-containing spray cans be labeled so that aerosol products that did not contain the chemicals would not be "unduly penalized."

The IMOS report was hardly the deathblow it could have been, but it nevertheless stunned the fluorocarbon and aerosol industry and provoked great cries of protest. One trade magazine, *House-*

hold and Personal Products, said it "dropped like a bombshell and left industry in disarray."

The IMOS report was widely interpreted as a verdict of "guilty until proved innocent." Certainly that was the way many in the industry took it. *Aerosol Age* said the IMOS report had all the clout of banning fluorocarbon-containing aerosols, but said the industry would not succumb.

Du Pont voiced "strong disagreement" with the IMOS recommendation concerning regulatory action, saying it was "tantamount to prejudging the results of the research. . . ." (An IMOS spokesman later said: "We do not see it that way. . . . We felt that it would be a disservice to the public and also the affected industries not to indicate at this time the way the evidence seems to be pointing.")

But Du Pont and others also took the tack of trying to interpret the IMOS report in the best possible light. One Du Pont release attempted to turn the tone of the IMOS report around 180 degrees: ". . . the information given the task force confirms that there is no imminent hazard to the ozone or to the public health from continued use of fluorocarbon propellants and refrigerants until the research programs have been completed. This fact should ease public concern." (Contrast this with IMOS' statement that there was "legitimate cause for concern.")

Du Pont scientist Richard Ward is quoted in an industry statement as saying that the IMOS report "essentially concurs" with the industry's position that there was no appreciable danger in continuing to use fluorocarbons while scientific studies were going on. Jack Dickinson of Gillette, then head of COAS, said that IMOS, "despite some language that has been misconstrued," essentially conceded the scientific uncertainties in the fluorocarbon/ozone hypothesis.

The IMOS report at least provided the congressional hearings with something new to argue about. And a good thing, too; by midsummer, with positions having hardened to the point of immobility, the hearings had begun to sound redundant.

In September 1975, Dale Bumpers' Senate Subcommittee on the Upper Atmosphere held a marathon session that stretched over seven days. The stratospheric debating society was out in force—some twenty-eight witnesses were heard. (Much of the testimony was highly technical, and at one point, the chairman muttered: "This is pretty tough on a guy that never even had high school chemistry and has practiced law all his life.")

The senators challenged industry's motives in appealing for a delay in regulations. Bumpers, although he said he was not referring specifically to Du Pont, said that Congress was "a little bit apprehensive about the good intentions of industry." He said there was a "normal inclination" to believe that industry was more concerned about economics than human health.

His witness at the time was Roy Schuyler of Du Pont's Organic Chemicals Department. A balding, jovial-looking type, Schuyler went all folksy. He revealed that he had four young grandchildren and "I would not want this advertised [he said before a public hearing of a congressional committee], but they call me, the two older ones, 'Papa Sweetie Pie.' Now, if I felt that there were any problem or danger, believe me, to the children down the way, I would want to pull the plug right now and be precipitous [sic]." But he firmly believed "there is no reason to ban now."

Another of the senators, Pete Domenici of New Mexico, wanted to know if Du Pont was spending more research money trying to disprove the fluorocarbon/ozone theory than in trying to find alternatives, particularly for refrigeration. Ray McCarthy estimated that Du Pont was spending five times more in the search for alternatives than it was spending in studying the Rowland/Molina ozone hypothesis. Schuyler added that Du Pont's Freon sales manager had been complaining that "we are not dedicating enough money to enhance the use of Freons in aerosols and other uses and I think that he is complaining because too much of our money is dedicated in the area that Dr. McCarthy has just identified."

The hearing closed with testimony from scientific experts. Sherry Rowland reviewed some of the latest data, including measurements that had been taken in the stratosphere, and argued that the theory was "no longer simply a deduction from chlorine measurements in the laboratory. . . . The situation as far as the experimental basis is very much more secure."

In large part, the urgency in Rowland's appeal for a ban was based on the time-lag problem inherent in the ozone/fluorocarbon theory. "The commitment we make now is for all of the next century," he once said. His reasoning went like this: It takes the fluorocarbons a long time to reach the stratosphere, break apart, and destroy ozone. Most of the chemicals already released were still in the reservoir of the lower atmosphere and had not even begun to do their damage in the stratosphere. With fluorocarbons being added to the reservoir year after year, the total impact on the ozone layer would ultimately be increased and the effects would last longer. According to this scenario, the sooner a ban was instituted, the better.

Senator Bumpers asked Rowland for an explanation of this concept, and when Rowland finished, Bumpers asked: "We have got a time bomb ticking up there, then?" Rowland played that one cautiously. "I don't want to overstate the case on this effect," he responded. He said the effects had already started and they would continue to build up, but it was not a situation that would suddenly explode twenty years hence.

It was inevitable that scientists at these hearings would be asked for their advice and opinions about regulating fluorocarbons. Some clearly felt a little odd playing this role. Bumpers asked Jim Anderson, a young University of Michigan researcher, whether the subcommittee should wait for more measurements to be made before recommending a ban on aerosols. When Anderson hesitated, Bumpers evoked laughter from the audience by saying: "I am not trying to put you in a box of any kind. Just give me a 'Yes' or 'No' answer."

"I do feel that the four walls are closing in," Anderson replied uneasily. But he said he favored cutting back seriously on the use of fluorocarbons in cosmetics, but not in refrigeration.

Mike McElroy played it even more cautiously. "I would rather decline to make any recommendations. My advice to you on that issue isn't worth any more than the advice of any informed layman. . . . It is my role, as I see it, to try to define the scientific issue."

As the hearings continued, it became increasingly clear that attitudes were hardening against the spray can. At Senator Bumpers' sub-committee hearing, the chairman challenged Jack Dickinson of

Gillette (who was also then head of COAS) with the argument that there were, after all, acceptable substitutes for all personal products. Dickinson responded that the overwhelming majority of consumers continued to prefer aerosols, and eliminating them would take away freedom of choice.

Bumpers immediately shot him down on that one. "You know, Mr. Dickinson, that argument is really not very persuasive to me at all, about what consumer preference is. I believe in consumer preference, but of course children have a preference for playing with fire, but we do our best to keep them from doing it. . . . In a case such as this, where the hypothesis can be so catastrophic, consumer preference should not be the guiding light. . . ."

Dickinson backed down. "No, sir. I don't think it should either. The health consideration comes first, but I respectfully submit that so much at this point is based on supposition and hypothesis that it is really inappropriate to act so precipitously."

As 1975 moved into 1976, there were numerous signs that the day of the fluorocarbon-containing spray cans were numbered. Arthur D. Little, Inc., a consulting firm that had done an economic study on the subject for the Environmental Protection Agency, was advising its clients that a ban should be anticipated within two to three years. The EPA suggested that industry should start looking for substitutes for fluorocarbon propellants in pesticide spray cans (over which the EPA had jurisdiction). The FDA had formally announced its willingness to institute a ban, pending soon-to-be-completed scientific studies.

If the handwriting was on the wall, however, the fluorocarbons had at least received a temporary stay of execution. A consensus was emerging that regulations would probably not come in before 1978. (Since the regulatory process could take some two years to complete anyway, the regulations could not become effective much before 1978, even if the agencies started immediately.)

Moreover, the flurry of activity at the state level had died down. Only two states, Oregon and New York, had actually passed legislation relating to fluorocarbons. In mid-June, Oregon had become the first political jurisdiction anywhere to pass a bill banning the sale of fluorocarbon-containing spray cans. The ban was to go into effect nearly two years later, in March 1977.[9] Debate on the bill had been lengthy and intense: Since Oregon's actions were precedent-setting, the pro- and anti-aerosol spokesmen were out in force. The Eugene *Register-Guard* reported that "a benumbed committee sat through 4½ hours of testimony with each member growing more bewildered by the scientific testimony of each successive witness."

In the end, the Oregon legislature passed the bill, and Governor Robert W. Straub, who had described the action as "consistent with Oregon's pioneering efforts" in promoting environmental quality, signed it.

Industry was predictably outraged. One industry spokesman described the action as a "multimillion-dollar false alarm." Another predicted in the Portland *Oregonian* that a "black market" would spring up—the bill, after all, prohibited the sale, but not the use, of the spray cans. "There's no question that when consumers are faced with a shortage of such household items as insect repellants or spray deodorants in Portland, they are not going to go back to fly swatters, Flit guns, roll-ons, or pads." Indeed, they would drive to Vancouver or Washington State "to get what they need." (It was undisclosed whether the FBI would be called in to deal with people transporting fluorocarbons across state lines.)

The Oregon action was unquestionably a setback for industry. It demonstrated, for example, that the spray can was inevitably becoming an albatross. As one Oregon legislator put it: "If we err on one side, we lose some spray cans. If we err on the other side, there's possibility we'll doom mankind."

The Oregon move was also discouraging to industry because they had lobbied at the state level almost as vigorously as they had at the national level. In fact, COAS had gone to considerable expense and effort to field witnesses during the state hearings. Industry seemed particularly dubious about state legislators, apparently regarding them either as political opportunists or ignoramuses who could not understand the scientific complexities of the issue.

The industry argued that state action was useless because the ozone problem was global in nature and it required a global solu-

[9] The ban went into effect on schedule. Under the act, shopkeepers selling fluorocarbon-containing spray cans were subject to a $1,000 fine or a year's imprisonment or both. The state accounted for about $40 million worth of spray cans annually.

tion. This was true enough, but the industry arguments carefully—probably knowingly—avoided the fact that positive measures by the states could generate a momentum and a legitimacy for a national fluorocarbon ban.[10]

However, the passage of the Oregon bill did not start a stampede. Although more than two dozen states (and even some municipalities) considered bills or bylaws to ban, restrict, or study fluorocarbons, many of these were quickly killed, and by the end of 1975, the rest lay dormant. New York was the only state besides Oregon to take definite action. It passed a bill, effective in April 1977, that required all fluorocarbon-containing spray cans to carry a label warning that they contained a propellant "which may harm the environment." It also permitted the state's commissioner of environmental conservation to impose a ban if he determined fluorocarbons to pose a hazard to health or the environment.

Perhaps the major factor in the slowdown of activity was the study being done by the National Academy of Sciences. A tacit consensus seemed to emerge that political and regulatory action should await the verdict of the Academy. The Consumer Product Safety Commission, for example, rejected petitions to ban fluorocarbons on the grounds that there was insufficient evidence that the fluorocarbon/ozone theory was correct and it was therefore impossible to determine that an "unreasonable risk of injury" would ensue. The CPSC decided to wait for the National Academy study. The Natural Resources Defense Council promptly petitioned the CPSC again to ban fluorocarbons and dismissed the Academy study by saying the scientists doing it "aren't on the front lines in this area anyway"—a statement that was demonstrably not true. Ten states joined the petition, including Florida, which expressed concern that the fear of increased skin cancer might scare tourists away from the "sunshine state."

But the fact was that nothing was going to move until the Academy rendered its decision. Even Sherry Rowland acknowledged that "at present, the argument 'wait for the NAS report' is almost un-

---

[10] Industry spokesmen occasionally even tried to suggest that action by the United States would have little impact because the United States produced only half of the world's fluorocarbons. While it was certainly true that international action was needed, it was ridiculous to suggest that a U.S. ban would not have a profound effect on the fate of fluorocarbons nearly everywhere.

beatable with legislatures." Or, as Senator Bumpers put it, "we are all waiting for the NAS to do their thing."

If any single organization can be said to embody the collective wisdom of American scientists, the National Academy of Sciences is it. The Academy has, of late, had its detractors and critics, but one can still hear it referred to in almost reverential tones, and it is certainly true that it essentially became the scientific Supreme Court in the case of the fluorocarbon debate.

The Academy would later be accused of procrastination on the fluorocarbon issue, but it actually got into the game right at the start—before the start, in fact. Bill Spindel, executive secretary of the NAS Chemistry Division, had heard about the Rowland and Molina work before the first paper was published in *Nature*. He went to speak to Rowland at the American Chemical Society meeting in Atlantic City in September 1974 and shortly thereafter got permission to assemble an ad hoc panel to advise the National Academy on what further action it should take.

Don Hunten from the Kitt Peak National Observatory was chosen to head the panel. Hunten, an expert in planetary atmospheres, had studied the problem of chlorine in the atmosphere of Venus. The other members of the panel were Sherry Rowland, Harold Johnston, Mike McElroy, and Francis Johnson of the Center for Advanced Studies at the University of Texas. The ad hoc panel met in Washington on October 26 and, at the end of the one-day session, it produced a brief report for the Academy. The Academy later released a short statement in the name of the panel, saying that the five scientists had concluded that, on the basis of existing evidence, "the problem is serious and will be acute unless the production of these substances is curtailed or stopped within a few years." The report acknowledged that some missing factor might show up and this, combined with the economic impact of an immediate ban, "suggests that a year should be allowed for detailed discussion. If no missing factors are turned up by then, drastic action will probably be necessary because the effects mount rapidly with time."

One recommendation was that the Academy should quickly issue the statement because of the "enormous public interest in the matter." The second recommendation was that the Academy form a committee to see that the stratospheric chlorine problem be "given broad, prompt attention" and that the Academy issue a final report within a year.

The activities of the ad hoc panel inevitably attracted press attention. Shortly after the panel met on October 26, Hunten was quoted in the New York *Times* as urging an immediate halt to the purchase of spray cans containing fluorocarbon propellants. Hunten, the story said, was speaking personally, not for the ad hoc panel. He refused to divulge the panel's conclusions, but said that, in view of the urgency of the situation, he hoped the assessment of the problem would be completed within a year.

In late November, the Washington *Post* ran its first major piece on the fluorocarbon problem. The story, by science writer Tom O'Toole, was pegged to the deliberations of the ad hoc panel, and it quoted Hunten saying that "my own personal feeling is that this hazard should be studied critically for another year, at which time action should be recommended. I think the action is obvious, that we should prohibit the manufacture of the stuff."

This, of course, immediately elicited a protest from industry spokesmen, who cried "fix!" Robert Ackerly, counsel for the Chemical Specialties Manufacturers Association, wrote in *Aerosol Age:* "One would assume that we are dealing with an objective panel of scientists without preconceived conclusions." He then equated Hunten's remarks with the sentiments in Lewis Carroll's famous poem: "I'll be the judge and I'll be the jury," said the cunning old Fury, "I'll try the whole case and condemn you to death." Irked, Hunten responded that his statement concerning a ban on fluorocarbons had been taken out of context. "The second sentence quoted should have been conditional: 'If the present indications are confirmed by this study, we will be faced with a serious future hazard. Action will then be required; I think the action is obvious, that the manufacture of fluorocarbons may have to be prohibited.' Unfortunately, I have no control over which words are chosen to be quoted by the newspapers."[11] Hunten said that name-calling would

11 However, Hunten can't get off that easily. He is also quoted in a United Press International story of November 27, saying: "My personal feelings are that it is really serious and drastic action is going to be necessary within a year or two to protect the ozone layer unless there is some completely new factor about the chemistry of the stratosphere that nobody has thought of yet. That is a remote possibility."

not help to establish the facts. He added that just because some environmental hazards had been overblown, it did not mean that this was always the case. Scientists had been searching "long, hard, and sincerely for an escape from the conclusion that chlorine atoms offer a serious hazard to the ozone layer. So far, the conclusion has only been strengthened."

Hunten concluded with an admonition: "Perhaps the shepherd boy is crying 'wolf' once again; but let us remember the end of the tale: The wolf did finally appear."

The furor in the media was not to the National Academy's liking and, though it had announced in mid-November that it would establish a formal panel to study the problems, it virtually vanished from the scene for the next four months.

During the four-month delay, the Academy was accused of foot-dragging by several people, including some members of the ad hoc panel, particularly its chairman, Don Hunten, and Harold Johnston. Ironically, there is good reason to suspect that the delay was caused in no small part by the Academy's wish to let the dust settle after the controversial press coverage that had attended the deliberations of the ad hoc panel.

At least one Academy official maintained that there wasn't really a delay; Bill Spindel said that things moved about as quickly as the mechanics of assembling the panel and arranging for funding would allow. Obtaining the money to do the study had not been particularly easy. The Academy had to solicit funds from the jealously guarded budgets of various Government departments and agencies, none of which were particularly anxious to part with the dollars necessary to bankroll the Academy's study; some agencies reportedly suggested that they might not be moved to do so unless and until they were legally forced to by the passage of bills pending in Congress that called for an Academy study.

Nevertheless, it was also true that the Academy had been upset with the publicity generated by the ad hoc panel. Its sensitivity to the matter of media coverage was made explicitly evident at the first meeting of the official panel on April 9, 1975, when the members were asked to refrain from discussing panel activities, particularly with the media. Such discussions, even of personal views, were deemed to be inappropriate—or what one panel member later referred to as "a hanging offense"—since, in the Academy's view, the panel was engaged in a "judicial review of the question." It was not even acceptable for members to speak only for themselves, for the press would inevitably identify them as members of the panel; this was, in fact, precisely what had happened with the ad hoc panel. "It's a specious kind of distinction," said Bill Spindel. "There really is no way for a member of the Academy panel to speak as an individual on that subject."

Nevertheless, some panel members, notably Fred Kaufman and Harold Schiff, protested that it was both unfair and unrealistic to expect them not to talk about the fluorocarbon issue at all. Several panel members were actively engaged in research in the field, and they fully expected to talk about their work at scientific meetings. Moreover, queries from the press were unavoidable and should not be brushed off with a "no comment." A compromise was struck: The panel members could discuss matters of fact but were to refrain from expressing personal opinions, particularly on the delicate matter of regulating or banning fluorocarbons.[12]

The Academy panel was to be convened under its Climatic Impact Committee (later the Committee on Impacts of Stratospheric Change). It is important, in light of later events, to understand the distinction between these two groups. The panel confined itself to scientific questions; it would evaluate existing data, recommend further research, and identify gaps in the existing information. The committee would incorporate these findings into its own report. This second report would deal with the consequences of ozone depletion and also focus on the public policy issues, making recommendations as to what should be done about the problem.

The Academy panel was unveiled, as it were, at an American Chemical Society meeting in Philadelphia on April 9, 1975. It consisted of twelve scientists, including one each from Canada, Britain, and West Germany. The chairman was Herb Gutowsky, director of the School of Chemical Sciences at the University of Illinois.

[12] The Academy sent a notice to each panel member asking if they had previously taken any public position that "in your judgment might appear to *other reasonable individuals* as compromising of your independence or judgment and, hence, in some measure prejudicial to the work of the committee on which you serve." (Emphasis the Academy's.)

None of the five members of the ad hoc panel were chosen to be on this more official panel. This did not bother Sherry Rowland, who acknowledged that the ad hoc panel had contained "too many district attorneys and not enough judges."

The Academy panel was directed to report one year hence, officially (and, some said, appropriately enough) on April 1, 1976.

It would not have been surprising if this panel had consisted entirely of scientists who previously had nothing to do with the ozone question. As we have noted before, there had been a tendency to choose such people so as to avoid charges of vested interest. In the case of its fluorocarbon panel, however, the Academy—wisely, it turned out—opted for a mixture. While some of the members had had no previous involvement in stratospheric chemistry, they were respected researchers in their own fields. Moreover, at least half the committee had a more than nodding acquaintance with the field, and several in fact had a direct and continuing research interest. Ironically enough, the panel would more often be accused of ignorance in stratospheric chemistry than it would be of vested interest.

The panel almost immediately came under pressure from various quarters to render its decision posthaste (this would steadily intensify as the regulatory agencies and the government departments—which were funding the study—came to realize that they could not proceed with their plans and research programs until the Academy made its report).

Some of the early pressure came from the old ad hoc committee members. Early in May, Don Hunten wrote to the panel to say that the NAS would "have a lot of explaining to do if a published assessment of the fluorocarbon issue was not released by the end of 1975." If the Academy did not intend to do this, it had better say so promptly and publicly "because at present the Academy is on record as intending a report this year."

The redoubtable Hal Johnston took issue with the whole concept of the panel. As far as he was concerned, the ad hoc panel had *not* recommended that the Academy set up yet another panel to investigate the problem. What they *had* urged was adequate financial backing for the researchers who could obtain the necessary data to resolve the question. The ad hoc panel, he said, believed that with such backing, the relevant data could be gathered within a year or two.

There seems to have been some misunderstanding between the Academy and the ad hoc panel on this score. According to Bill Spindel, "no one charged the (ad hoc) panel with preparing an Academy report. . . . The Academy is not going to issue a one-page statement prepared at the end of a four-hour meeting as its statement." The function of the panel, he said, was to provide advice for further study by the Academy, and their report was strictly an internal document.

However, Hal Johnston seemed less irritated by this point than by the fact that, in his view, everyone was ignoring what had been accomplished in the three-year CIAP program. In response to a request for comments on the Academy's plans, he wrote a letter that characterized the members of the Academy's panel[13] as scientific innocents who were almost bound to be taken in by the wily strategists of the chemical industry. Some in the industry, he said, were engaged in "a self-serving campaign of deception, false statements, omissions, 'red herrings,' and subtle character assassination directed against Rowland and others. Some assert that the chlorine-catalyzed destruction of ozone is 'just a theory' inspired by irresponsible 'computer jockeys' and that virtually nothing is understood about the case. . . . They propose slowly to spend $3 million to $5 million over the next three years, but they close their eyes to the $20 million to $50 million spent on stratospheric research during the past three years."

Johnston clearly did not think much of the panel's ability to cope with this clever campaign. In his letter, which was hardly calculated to endear him to the panel, he noted that its members consisted primarily of scientists who were not experts in the specific problems of ozone depletion by NO_x and chlorine catalysts and that, with some exceptions, "the members of your committee are ig-

[13] Johnston actually uses the term "committee," but the tone of the letter indicates that he is talking about the Academy's scientific panel. For one thing, he says he is expanding on remarks he had made at the Philadelphia ACS meeting; it was at that meeting that the panel's membership was announced. Johnston's letter also deals with scientific details that were the purview of the panel, not the committee. Finally, at the time the letter was written, the membership of the committee had not been announced.

norant of the nature, extent, and pertinence of the CIAP findings."
In consequence, the panel members were highly susceptible [to] red
herrings. "You may believe the chemical industry spokesmen who
say 'nothing is known' because some of you don't know very much
yourselves. You accept as reasonable that it will take three to five
years to get the asked-for answers because you don't understand
how very much was learned during the last three years. . . ."

Johnston acknowledged that the letter was blunt—he said he had
previously stated his position in polite terms but "in case the polite-
ness concealed my meaning, I will go out of my way to be blunt
here"—and he apologized for it at the end. Nevertheless, the letter
understandably annoyed the panel members. Five of them—Julius
Chang, Robert Dickinson, Fred Kaufman, James Friend, and
Harold Schiff—had been directly involved in the CIAP study.
Dieter Ehhalt had made many crucial measurements of stratospheric constituents, and his data were used in the CIAP study.
Brian Thrush, the British member, was certainly familiar with the
relevant chemistry. Finally, George Pimental was a close associate of
Harold Johnston at Berkeley and was up to date on the ozone issue,
though he was not directly involved in it. Johnston's charge, if indeed it referred to the panel, was patently unfair. But it was not
without its amusing aspects. In the letter, Johnston referred to the
five members of the original ad hoc panel as "perhaps five out of
seven of the most knowledgeable experts on this subject." Schiff
thought this choice of numbers was a stroke of genius. As he
remarked at the panel meeting where Johnston's letter was
discussed: "Each of us is sitting here wondering who the seventh expert is." When told this later, Johnston did not get the joke. There
is a certain humorlessness in his approach to the ozone issue, and the
comment in his letter was clearly a case of unconscious wit.

About two months into their deliberations, the members of the
Academy panel got quite a shock. In mid-June, the IMOS report
was released, and the panel suddenly found itself charged with
doing an "in-depth" study. They had actually been planning to do
something along the lines of the IMOS report itself, but now they
had been pre-empted; not only that, they'd also been given a much
more ambitious mandate than they had originally anticipated. The

IMOS report had the effect of making the panel the Supreme Court on the fluorocarbon controversy.

In light of these developments, the panel members decided that the greatest contribution they could make would be to assess the degree of uncertainty in the calculations of ozone depletion. In addition, they decided to re-examine all aspects of the problem. How good were the computer models? Did laboratory and atmospheric measurements support the theory? Had anything been overlooked that would make the problem much more or much less serious than scientists believed? Were there any measurements that could prove the case beyond a reasonable doubt? In short, was there—to use the language of Watergate—a "smoking gun"?

CHAPTER NINE

The Search for the Smoking Gun

The damn stratosphere was too messy a place to fool around with.

—TOM DONAHUE, in an interview, November 1976.

The scientific battle over fluorocarbons revolved around a single issue: the question of proof—proof that the scenario painted by Rowland and Molina was actually taking place in the stratosphere. Obtaining such proof required a detailed understanding of what the stratosphere was like—an understanding that was by no means in hand when the ozone controversy arose at the beginning of the 1970s. At that time, the stratosphere was still largely a no-man's-land. In fact, it had been dubbed the "ignorosphere" because aeronomers had paid so little attention to it in the past.

This was so for a number of reasons. Meteorologists were concerned with the lower atmosphere, where the weather occurred. They studied atmospheric dynamics and transport—the movement of large air masses—and were generally not concerned about the role of chemical reactions. The stratosphere was uninteresting to them—there was no weather to speak of up there, and meteorologists were not particularly intrigued by chemical processes, which they considered to be relatively less important in the lower atmosphere than in the stratosphere.

The aeronomers also avoided the stratosphere. In the early days, basic research in aeronomy was supported mainly by the Department of Defense (DOD), which was interested in the phenomena associated with the re-entry of missiles into the earth's atmosphere. Since these phenomena largely occurred in the regions of the atmosphere above the stratosphere—called the mesosphere—aeronomers simply neglected the stratosphere. For one thing, the chemistry of the mesosphere was much simpler, and transport was relatively unimportant. For another, "we had troubles enough with the part of the atmosphere we were being paid to study by the DOD," said Donahue.

Also, after the Second World War, scientists were able to use rockets, which carried their instruments higher into the atmosphere than they had ever gone before. Anxious to play with these new toys, the scientists were not interested in lower regions of the atmosphere, which had been accessible for some time with balloons.

But the Defense Department more or less abandoned the aeronomers in the 1960s. With a ban on atmospheric testing of nuclear weapons and with the United States and the Soviet Union seemingly at a nuclear standoff insofar as ICBMs were concerned, the urgency for basic research on the upper atmosphere, at least on the part of the Defense Department, had diminished considerably.

The aeronomers looked around for a new home, and soon they found it—at NASA. The space agency was sending unmanned probes to other planets in the solar system, and the aeronomers figuratively rushed happily off to Mars and Venus to study the exotic atmospheres there. Nor did this immediately prompt them to turn their attention back to the stratosphere of their own planet. The stratosphere was unappealing in those days, says Donahue, because it was "hopelessly messy chemically." Moreover, scientists would not only have to sort out this chemistry, but also figure out how it interacted with transport processes (air motions) to influence stratospheric conditions—a chore that even today is perhaps the major source of headaches in stratospheric science. It all hardly seemed worth the effort, particularly when, as Donahue put it, "we had so many other interesting things connected with Mars and Venus."

Nor were laboratory chemists up on the chemical reactions going on in the stratosphere. Their work was geared to providing data that were of interest to the aeronomers, who still had their heads in the mesosphere because that was where the action was.

This has frequently been advanced as the reason why the scientists who might logically have been expected to identify the man-made threats to the ozone layer did not in fact do so. But whatever the reasons, the fact was that the scientific understanding of the earth's stratosphere and its chemical processes was wholly inadequate to assess the situation when the SST was identified as the first in what would become a long list of threats to the ozone layer. In an unpublished manuscript about the SST controversy, chemist Ian Clark has drawn a chart showing the level of scientific activity (man-hours per year) devoted to studying various regions of the atmosphere. The low point in expertise approximately coincides with the region of maximum ozone concentration and the proposed altitude of SST flights.

As we have seen, however, the recognition of the SST threat galvanized the scientific community into action. Donahue correctly points out that their other endeavors gave aeronomers an important base of expertise and enabled them to switch their attention quickly and efficiently to the SST problem once it had been identified. As a result, the scientific community was caught less off-guard on the subsequent fluorocarbon issue; the CIAP program had, after all, given them a three-year head start. The stratosphere was, by that time, if perhaps not absolutely tidy, at least less "messy" than it had been. Scientific understanding of stratospheric chemistry had been vastly improved, as had the computer models used for predicting changes in the earth's ozone layer. Moreover, CIAP had also set in place a network of scientists who could quickly swing into action to tackle the fluorocarbon problem.

But before examining the scientific quest for proof of the Rowland/Molina theory, it is important to discuss the "ozone myths." These were a number of arguments put forth by industry (and others, including some scientists who should have known better) as persuasive evidence against theories that human technology could destroy the ozone layer. These myths were characterized both by their persistence (most of them had also been raised during the SST controversy) and by the misunderstandings that accompanied them.

The first myth goes like this: If the ozone layer varies so much naturally, we shouldn't worry about the possibility that we might reduce it by, say, 10 per cent.

There is no question that the ozone layer is indeed highly variable; it changes not only from place to place, but also from time to time. Ozone is produced primarily over the equator, but it is steadily transported toward the poles. Thus the higher latitudes always have a thicker ozone layer. In fact, the highest concentration of ozone at any time occurs over the North Pole in early spring, when there is little sunlight there (ironical, in view of the fact that ozone is largely produced by the action of sunlight over the equator).

There is little seasonal variation in ozone at the equator, but as one proceeds toward the poles, the variation becomes increasingly pronounced. At very high latitudes, there can be a difference of 200 per cent between the maximum in the spring and the minimum in the fall. Over the continental United States, there is a variation of 20 per cent between spring and fall. The ozone also varies on a daily or weekly basis, primarily because of air motions that move it around. Over a decade, a given location will experience about a 10 per cent variation in ozone.

Measurements of ozone have been made with increasing accuracy and sophistication since 1929, but even with modern technology, it is still difficult to measure exactly what is happening to the ozone layer. More than any other single factor, this great natural variability has thwarted the efforts of scientists to determine exactly what human technology is doing to the ozone layer. The predicted effects from human activity—usually in the range of a few per cent, although sometimes higher—were generally much smaller than the natural variability, which could range, as we have seen, from 10 to 200 per cent. Certainly any effects of human technology to date have been lost in the "noise" of the natural variations.

The important distinction to make here is that there is a vast difference between large variations that occur in a given location over a short time frame and changes that occur in the *annual average amount of ozone over the entire earth*. In the latter case, a change of a few per cent can be significant. An analogy with temperature makes the distinction clearer: A difference of five degrees, or even more, between one day and the next in a single location is neither unusual nor difficult to cope with. (In fact, much larger differences often occur between day and night in the same location.) A difference of five degrees in the *annual average temperature* of the entire earth is the difference between the present climate and full ice age conditions or, alternatively, a five-degree change in the opposite direction is the difference between the present and the melting of the polar ice, which would probably put many coastal areas under water.

There is a wrinkle to the ozone-variability myth that has been much favored by the aerosol and SST industries. It goes like this: This ozone depletion caused by the SSTs or spray cans cannot be harmful because people experience even greater variations by moving from, say, the northern part of the United States to the South. (As we saw in Chapter Two, the sun's rays slant through less ozone toward the equator, and therefore a greater amount of ultraviolet reaches the ground there.)

Here again, the temperature analogy is useful. The average annual temperature in Houston is 69 degrees F., while the average annual temperature in Chicago is 51 degrees. If you move from Houston to Chicago, you voluntarily choose to live in a climate that is, on average, 18 degrees cooler. But it is a very different story to suggest that the *whole world* be cooled down by 18 degrees; not only would *everyone* be subjected to a cooler climate whether they chose to be or not, but the effect on agriculture and the whole ecosystem would undoubtedly be devastating.

This is the crucial point about ozone depletion—that one is talking about the *average reduction over the entire earth*. This is not analagous to someone simply moving closer to the equator. Everyplace on earth would be subjected to higher levels of ultraviolet radiation and there would, for example, be an absolute increase in skin cancer unless *everyone* moved to a higher latitude. Even if you were to ignore the practical and political difficulties of doing this with four billion people, the fact remains that plants, animals, and crops cannot so readily be shifted around.

National Academy studies noted that living things have also adapted to their climate and soils, which also vary from place to place. Migration to higher latitudes to avoid increases in UV-B "would require readaptation in many other ways as well." They also pointed out that plants seem to cope with as much as a 20 per cent change in UV-B (the result of the 10 per cent ozone variation over periods of a decade). But ozone reductions caused by human activities would be superimposed on the natural variations and would intensify them. "During periods of low ozone concentration, biological systems would be exposed to levels of UV-B higher than ever before experienced."

The ozone variability argument was further complicated in mid-1974 by a report published in *Science* magazine by Julius London and Jean Kelly of the University of Colorado. London and Kelly reported that during the 1960s, there had been an increase in ozone of about 7.5 per cent in the Northern Hemisphere.

This was immediately seized upon by many industry spokesmen. Typical of the industry argument was a statement made by Ray McCarthy in congressional testimony in December 1974: ". . . there is no experimental evidence supporting the chlorine/ozone theory. To the contrary, a study by London and Kelly indicates that the concentration of stratospheric ozone has actually increased during the past decade." Ralph Engel of the Chemical Specialities Manufacturers Association added that this increase in ozone coincided with the period of peak fluorocarbon production.

Frequently implicit in these citations of the ozone data was the argument that they were actually inconsistent with the Rowland/Molina theory. (In fact, some industry spokesmen who were not scientists explicitly claimed this.) But the argument is flawed in several respects. First, it ignores the time-delay problem in the fluorocarbon/ozone theory. While the 1960s represented the period of peak fluorocarbon production, most of the chemicals were still in the lower atmosphere and had not even begun to damage ozone; second, there was no way of telling that the increase in ozone would not have been *greater* if the fluorocarbons had not been released to the atmosphere; finally, scientists just didn't know what caused this increase, although there were a lot of theories. In the end, however, scientists concluded that the increase was part of the natural ozone cycle. In addition to the weekly and seasonal variations already discussed, ozone may vary with the eleven-year solar cycle, and there may even be longer, hundred-year cycles. We don't really understand these trends, and to gain a better understanding would require several centuries of observations. This is impractical for political decision-making and is the main reason why arguing about ozone measurements and what they mean is not a fruitful way to solve the immediate problems.

Nevertheless, early in the fluorocarbon debate, industry persistently raised the ozone data to bolster their case. Some seemed to want it both ways. On the one hand, they would argue that the great variability of ozone meant that man-made effects were neither detectable (which was true) nor important (which was not necessarily true)—or that scientists simply didn't understand the complexities of the ozone layer. On the other hand, many seemed quite prepared to accept the observed increase in ozone as proof that human technology was *not* having an effect.

(This willingness to extrapolate from the available data was shared by the SST industry. In Chapter Three, we saw that Arnold Goldburg, chief scientist for the Boeing SST, argued that Jim McDonald's ozone/SST theory was all wrong because ozone and water vapor had both been increasing when the theory suggested that ozone should be going up while water vapor was going down. But the water-vapor measurements had been made in one location only; they were not global figures, and ozone variations are so complicated that it is impossible to draw conclusions from such isolated observations.)

This myth has become largely moot with recent evidence that indicates that the ozone trend has reversed and the ozone levels in the 1970s have been dropping. The changes in both directions have been comparable to historical ozone variations, and no one can claim to know exactly what causes them or why they reverse. It would be just as wrong to condemn the fluorocarbons on the "evidence" of current declining ozone levels as it was to exonerate them on the "evidence" of increasing ozone levels in the previous decade.[1]

However, after making so much of the increasing ozone levels, industry had to come to terms with the new data indicating a reversal of that trend. Here again, some seemed to want things both ways. For example, at one hearing, Robert Orfeo, a research director with Allied Chemical, attributed both the increasing and the decreasing ozone trends to natural cycles. He does, however, single out the "much-discussed decline in ozone level observed for the first half

[1] Although the ozone data are not good enough for immediate political decision-making purposes, scientists are continuing their attempts to build up a worldwide ozone-monitoring system that will enable them to provide better data in the future.

of the 1970s" as "a predictable phenomenon from historical patterns. . . ." In other words, the decline was natural and not attributable to fluorocarbons. Fair enough. Yet, not two paragraphs later, he again raises the familiar industry argument that implicitly exonerates fluorocarbons, by saying that analysis of the ozone data "indicates that the ozone level in the stratosphere actually increased in the 1960s, a period of high production of chlorine and chlorine-containing products."

Some in industry continued to insist, on the basis of the ozone increase, that the ozone shield was never healthier, but by mid-1975, after the declining ozone trend became established, many industry spokesmen simply abandoned the argument. (Richard Scorer, on the other hand, did not. In June, he criticized a newspaper editorial that mentioned the ozone drop as "just absolutely wrong," insisting that ozone had been increasing for the past few years.)

The second myth has to do with the "self-healing" effect of ozone. The argument goes like this: If ozone is destroyed, then more ultraviolet radiation will get down to lower layers of the stratosphere. There it will simply be absorbed by oxygen molecules, which will split apart into atoms that will then join with other oxygen molecules to form more ozone. The net effect would be a lowering of the altitude where the maximum ozone concentration is located; but since we only care about the total amount of ozone over our heads, this doesn't matter.

There is some truth to this scenario, but it is only partly correct. Oxygen molecules and ozone do not absorb the same part of the ultraviolet radiation, so a significant portion of the UV that gets through the depleted ozone high in the stratosphere will *not* be absorbed by oxygen molecules lower down. Instead, much of it will reach the ground as UV-B.

The part of the ultraviolet that is absorbed by oxygen molecules will indeed make more ozone, so there is some self-healing. But it is not complete; for every molecule destroyed in the high stratosphere, you do not get another manufactured lower down. In any event, all the computer models used to predict ozone depletion from human activities have taken this limited self-healing effect into account.

Finally, suppose that there was 100 per cent self-healing. There would nevertheless be a gross redistribution of ozone that would upset the heat balance of the stratosphere and might cause climatic problems.[2]

The final myth might be called the "blind faith" myth. It goes something like this: Nature has been coping with ozone-destroying processes for billions of years. There are feedback systems, self-righting mechanisms, that enable the atmosphere to continue to cope. Human activities are far too puny to seriously harm such a large and resilient system as the earth's atmosphere.

Richard Scorer was one of the most outspoken proponents of this view. During his U.S. tour, he called the atmosphere "the most robust and dynamic element in the environment. Man's activities have very little impact on it." Scorer reportedly said that it would be difficult for fluorocarbons to do much damage that nature wouldn't correct.

This was a philosophy shared by many in the fluorocarbon industry itself. Ray McCarthy, for example, is quoted in one magazine article as saying, "I have faith in the biosphere acting to preserve life." He was, and is, a firm believer in feedback loops that work to keep everything in balance, and once suggested that fluorocarbons might be handled naturally in due course. McCarthy, who has a background in electronics, says that electronic systems seem to be able to cope with variations of up to about 10 per cent, but beyond that, feedback mechanisms cannot bring the system back to normal. He seems to feel that this is a universal law and that this tolerance range would also apply in the case of biological systems.

It is possible, of course, to get into endless philosophical debates about nature's ability to withstand the impact of human activities. There are genuine disagreements on this point, not only in the fluorocarbon debate but also in many other environmental controversies as well, Still, the faith of the fluorocarbon industry in nature's resiliency seems a shade too optimistic. Nature does need an occasional respite from the onslaught of human pollution; there are no infinite garbage cans. Sherry Rowland told *Food and Drug*

[2] For an explanation of this problem, see Chapter Ten.

Packaging magazine that "some people have the belief that the Lord designed an atmosphere that can take this kind of abuse. I live too close to Los Angeles to agree with that."

Los Angeles is, of course, a limited geographical area, and there are people who concede that humans can have a large local impact while believing that human activities are too insignificant to have a global impact. In the context of this argument, the ozone problem is a particularly interesting case study, because of the catalytic chain reactions. The existence of these chains suggests that even comparatively small amounts of pollutants can have a devastating global effect. It is only half of the story to argue (as some fluorocarbon industry officials have done) that there are only a few parts per trillion of fluorocarbons in the atmosphere and that a part per trillion is about one drop in a thousand large tankcars. The other half of the story is understanding what those few parts per trillion will do to the atmosphere.

Ralph Cicerone believes that the ozone controversy has played an important role in demonstrating this concept to the public. The concern over catalytic chain reactions has shown the public that the earth's atmosphere and oceans are not infinite renewable resources and that we don't have *carte blanche* to "put out as many pounds of junk as there are pounds of water or air to foul up."

It is at this point that people usually ask: "Why don't we make more ozone? If we can have a global effect in destroying ozone, why can we not simply counteract that effect?" Unfortunately, there are no catalytic chain reactions that make ozone. Ozone production involves a massive input of energy from the sun acting on a major constituent of the atmosphere, oxygen. Unlike the destruction of ozone, the production of ozone *is* too big a job for us.

There is one final point to be made on the "blind faith" myth. It involves a fallacy that stems from human arrogance—from the assumption that, to whatever extent nature *does* right itself, it will do so in a manner that protects human beings. What reason is there to believe that nature has any vested interest in preserving homo sapiens? Nature may indeed move to right itself; it may even, as McCarthy suggested, learn to "handle" fluorocarbons. But it may do so by reaching a new equilibrium, by having less ozone and, incidentally, by removing the irritant, man, in the process. And someone else, better adapted, will inherit what remains of the earth.

The ozone myths provided a persistent backdrop against which the fluorocarbon debate unfolded—and it must be said that the scientific community generally did a poor job in countering these myths in public debate—but, of course, the scientific part of the debate was much more specific than this. In large part, this debate focused on the single most difficult task—the task of finding proof that the Rowland/Molina theory was correct. As we have seen, industry's main argument was that it was "just a theory," a hypothesis fraught with assumptions and unsupported by experimental evidence. A recurrent theme was that no regulatory action should be taken until, as one Du Pont brochure put it, "research is completed." But when would the research be "completed"? When proof was found? The problem was that industry spokesmen seemed at times to have a rather flexible definition of what constituted proof. They outlined many things that had to be demonstrated before they would accept the theory, but it was impossible to pin them down to an admission of what they would regard as absolute proof—or what came to be known as the "smoking gun." If industry believed there was a "smoking gun," they weren't telling others what they thought it was. "That's a tough one, really a tough one," temporized Bunker Crawford of Du Pont of Canada Ltd. in response to a question about what he would accept as proof. "I don't think that's a question I can answer as an individual." Pressed on the point, he added: "I know at what point I would recommend we get out of it." He refused, however, to disclose what that point was. "I don't know how you'd ever get anyone to answer that question." Asked why not, he said: "That would be tantamount to slitting their throats."

However, Peter Jesson of the U. S. Du Pont company did say, in response to the same question, "I don't think we will require proof that is unattainable."

The most direct proof of the theory, of course, would be to continue using fluorocarbons and wait to see if the ozone layer starts to decrease. But there are several drawbacks to this method. First, the great natural variability of ozone would mask the effect for a considerable time, and it would be quite a while before you could be sure that anything was happening to the ozone. Moreover, you could never be sure that any changes you do observe were caused by the fluorocarbons. Finally, if you were to take action only when you could be sure that an ozone depletion due to fluorocarbons is observed, the ozone layer will not immediately recover. In fact, the depletion will get much worse for some time, and it will take an even longer time to recover to the level at which you first observed it. (This would be a much more serious factor in the case of fluorocarbons than in the case of SSTs, because fluorocarbons build up a reservoir in the lower atmosphere that can continue to contribute to ozone destruction for many years. SSTs on the other hand, inject their pollution right into the stratosphere; it stops as soon as the planes cease flying, and recovery of the ozone layer begins immediately.)

Since we cannot adopt a "wait and see" strategy, the next best thing is to *predict* what will happen if we continue to use fluorocarbons. This is where computer simulations come in. First, scientists try to represent the atmosphere as it is now. Their mathematical equations describe how materials put into the atmosphere distribute themselves around the globe, east and west, north and south, and vertically up through the atmosphere. This is called transport. Second, the equations describe how these materials react chemically with all the different kinds of molecules in the atmosphere. This description of the present atmosphere is called a computer model.

The computer is then instructed to project forward in time. It is given various scenarios for the amount of fluorocarbons being released (for example, fluorocarbon releases continuing to grow at 10 per cent a year, or, alternatively, a complete ban on fluorocarbons at some given date) and asked to predict how the ozone layer will change as a result.

The obvious questions are: How good are the models? Can we believe their predictions? There's a saying familiar to all computer scientists known as "garbage in, garbage out," and in order for us to put our faith in the computer predictions, the models need data somewhat better than "garbage in." This was perhaps the major task the National Academy panel set for itself—determining the quality of the data that went into the computer models.

The first important datum is the rate at which chlorine is being put into the atmosphere at present, both from natural and man-made sources.

Natural sources include HCl from volcanic eruptions and sodium chloride (salt) from sea spray. But these compounds are released mostly in the lower atmosphere, and both are highly soluble, so they are usually rained out of the atmosphere before reaching the stratosphere. Some volcanic eruptions, however, are violent enough to propel HCl into the stratosphere; this is what Cicerone and Stolarski talked about at Kyoto.

(Industry's attempt to make more of this fact than was really justified resulted in one of their more antic ventures, the "Waiting for the Volcano to Blow" follies. In October 1975, at an industry press conference, Jim Lodge of COAS said at a press conference that a volcano off the coast of Alaska was due to erupt; he and the industry were arguing that if the volcano spewed chlorine into the stratosphere and if the theory was right, there should be an observable reduction in ozone. If there wasn't, this would imply that the stratosphere was able to withstand injections of large amounts of chlorine and that fears for its integrity were unfounded. Donald Davis was characteristically skeptical of this heavy reliance on the volcano, which he regarded as "another blunder" by industry. The chances that it would produce conclusive ammunition for industry's case was "only fair," and even if the volcanic chlorine did get into the stratosphere, "will it necessarily disprove the Rowland-Molina contention that fluorocarbons are the principal villain?" Davis's misgivings proved founded. The volcano erupted near the end of January 1976, but the industry was evidently not at all happy with the results, for they did not call another press conference and, in fact, maintained silence on the subject for some months, finally stating only that the volcano had exploded prematurely and that the results were inconclusive in so far as the ozone theory was concerned.)

Volcanos aside, there are other natural sources of chlorine. One, discovered by Jim Lovelock, is methyl chloride, attributable mainly to kelp and seaweed. (Lovelock has also suggested that forest fires and slash-and-burn agricultural techniques might also produce significant quantities of methyl chloride.) Lovelock initially argued that methyl chloride was the dominant source of chlorine in the stratosphere and that fluorocarbons contributed no more than about 20 per cent. However, it is now known that about 90 per cent of the methyl chloride is actually removed in the lower atmosphere by chemical reactions, and only a small proportion of it gets into the stratosphere. In the mid-1970s, methyl chloride contributed about as much to ozone destruction as the fluorocarbons did at that time.

The fluorocarbons were, by that time, the largest man-made source of chlorine going into the stratosphere. It was relatively easy to obtain figures for the total U.S. production of these chemicals because such statistics were collected by the Tariff Commission. In addition, the fluorocarbon producers in the United States and in many other countries were very co-operative in supplying information on worldwide production. It proved impossible to get production figures from the Soviet Union and other Eastern Bloc countries, but it was estimated that these countries contributed less than 5 per cent to the total. (Figures for the Soviet Union are now available.)

Up to 1975, a total of some 28 billion pounds of fluorocarbons had been made, and some 90 per cent of them had been released to the atmosphere. Less than 5 per cent of the total amount produced has reached the stratosphere. Up to 1973, the United States produced more than half of the world's fluorocarbons, but after that it dropped to less than half.

Chlorine-containing substances other than fluorocarbons have also been manufactured and released in large quantities. Most of these have been used as solvents for dry cleaning and for metal degreasing. They pose less of a problem than fluorocarbons because they are largely removed by chemical reactions in the lower atmosphere, and only a small amount makes it to the stratosphere. These chemicals may, however, require scrutiny if their use continues to grow.

Carbon tetrachloride, which was once widely used as a cleaning solvent, is also present in the atmosphere. Like fluorocarbons, it is inert, and it makes its way to the stratosphere, where it contributes to ozone destruction. However, the use of carbon tet was restricted in the late 1950s and early 1960s for toxicological reasons, and its concentrations in the stratosphere are already declining.

Fluorocarbon industry spokesmen often tried to pass the buck, or at least part of it, to other man-made chlorine sources. It is uncertain what they hoped to accomplish by this; the tactic might well have forced the rest of the chemical industry to share the misery, but it would have done little to get the fluorocarbons off the hook. One does not make a problem go away by arguing that other things make the problem worse. The logical end point in this case is that all the trouble-making chemicals should be banned, not that fluorocarbons should not be.

The argument over the role of natural sources of chlorine was, however, a different matter. Here there was a prospect that fluorocarbons could find a reprieve.

Industry representatives frequently argued that there were large natural sources of chlorine; HCl from volcanos and methyl chloride were the two most widely used examples. (One Du Pont pamphlet categorically included slash-and-burn agriculture among the things that contribute large amounts of chlorine to the atmosphere, even though, at the time, this was "just a theory.") The industry generally had two points to make. The first was that these large natural sources demonstrated that the earth's atmosphere had been coping nicely with chlorine for billions of years. The second was that the natural sources might swamp the fluorocarbons, which could not then be considered a problem. Ray McCarthy argued, for example, that even in a quiet year, volcanos produce five to ten times more chlorine than fluorocarbons. "If the natural background of chlorines is large enough," said a Du Pont brochure, "the addition of fluorocarbons would have an insignificant effect on ozone."

There are many things that can be said about this. The first is that during those billions of years when the atmosphere was "coping" with natural chlorine, life on this planet was evolving to "cope" with the resultant conditions. What we are concerned about now are relatively sudden *changes* to the natural situation. That brings us to an important second point. Most of the industry statements carefully talk about natural sources of chlorine in the *atmosphere*, not the stratosphere. The distinction is important. It is the amount of natural chlorine in the stratosphere—and how much man-made chlorine will add to it—that matters. The statement in the Du Pont brochure should have read: "If the natural background of chlorines *in the stratosphere* is large enough, the addition of fluorocarbons would have an insignificant effect on ozone." (As we saw in Chapter Three, when Hal Johnston first suggested that NO_x from SSTs might be a problem, the first problem facing the scientific community was to determine how much natural NO_x was already up there and how much the NO_x from the SSTs would add to it. At the time they didn't know the answer to that question; that's part of what CIAP was all about.)

The best current data show that the stratosphere does not contain large amounts of natural chlorine. Most of the natural methyl chloride and the HCl from volcanos never makes it up there. In fact, the amount of chlorine that human technology has been injecting into the stratosphere in the 1960s and 1970s is already about *equal* to the natural sources of chlorine. Moreover, methyl chloride, for example, is being released into the atmosphere at a constant rate; the amount of fluorocarbons has been steadily increasing in the stratosphere, and this increase will continue for a time even if the use of fluorocarbons ceases, because of that insidious reservoir of the chemicals building up in the lower atmosphere.

Computer calculations have indicated that, for a constant input of fluorocarbons, the chlorine from human activities would eventually exceed natural chlorine in the stratosphere. This would be a gross disruption of the natural state of affairs, and it is *not* a situation the earth's atmosphere has been coping with for billions of years.

As we have seen, the first step in understanding the fluorocarbon problem is to determine the rate at which these chemicals, as well as other sources of chlorine, are being released into the atmosphere. The next step is to determine what happens to them. There are two facets to this problem: the familiar questions of transport (how fluorocarbons move through the atmosphere) and chemistry (how chemical reactions affect their fate).

The fluorocarbons sprayed from billions of cans over the years become distributed around the earth. The computer models attempt to describe these movements, but, unfortunately, they are not yet good enough to do this exactly. (If they could, we would have perfect long-range weather forecasting, for similar computer models are used in making such forecasts.) So approximations are necessary. Scientists know, for example, that air mixes more rapidly in the east–west direction than it does in the north–south direction, and that it does both more rapidly than it moves upward into the stratosphere. Thus most computer models assume an average value —that is, a uniform mixing—of fluorocarbons (and other molecules as well) over latitude and longitude and consider only how the fluorocarbons move upward with time. This is usually determined by measurements of how trace substances—such as radioactive debris from an atmospheric nuclear explosion—are distributed throughout the atmosphere. The use of this approximate method of portraying the atmosphere was sharply criticized by industry for being far too oversimplified to describe such a complex system, but industry nevertheless used the technique in some of their own studies.

The National Academy panel concluded that this treatment of the transport problem introduced an uncertainty of about a factor of two into the predictions of ozone depletion.

The next step in the process is to determine what happens to the fluorocarbons once they get into the stratosphere. How fast are they broken down by the ultraviolet radiation of the sun? What happens to the resulting chemical fragments? How do they affect ozone? What else happens to them? How quickly do these various processes occur? The task of answering these questions fell to the atmospheric chemists.

Fortunately, the laws of chemistry are universal, so the kind of reactions that occur, and the speed at which they occur, can be studied in the laboratory and the results used directly in the computer models. As we have seen, Tom Donahue described the stratosphere as chemically "messy," and there are, indeed, many coupled, competing, interacting, simultaneous chemical reactions that go on up there. Most models include at least eighty of these reactions.

The speeds (rates) of these reactions can be measured in the lab, but not with total accuracy. Some reactions are more difficult to measure than others and are therefore more uncertain than others. These uncertainties are compounded, and the final conclusion reached by the computer will reflect these cumulative uncertainties. Fortunately, however, not all eighty reactions are equally important.

Only about ten are of major importance, and a great effort was made to improve the measurement of these. In consequence, there was a period in 1975 when new data on the rate constants were shifting the predictions of ozone depletion back and forth, and this was the cause of yet another confrontation between the fluorocarbon industry and the scientific community. The most famous of these might be called the 300 per cent solution.

A full-page ad by Du Pont in many newspapers and magazines across the country was only one of many industry statements to claim that new laboratory data on the speeds of chemical reactions had substantially reduced the original predictions of ozone depletion. "In fact, the impact was overstated by 300 per cent."

This figure was culled from congressional testimony given by Ralph Cicerone. In describing lab tests done at the University of Maryland and the University of Pittsburgh, Cicerone had indeed said that these data meant that ozone-loss projections through early 1975 were too large by about 300 per cent (or a factor of three). But he protested that industry was making misleading use of this information by ignoring the fact that the uncertainties were working in both directions; while some of the new chemical-reaction rates did reduce the predicted impact, others had carried the predictions back in the other direction. "The net result is that the predictions now stand about where they were a year ago, only with more evidence behind them," he said. "The potential seriousness of the problem remains."

Since the point was raised in congressional testimony by industry, Cicerone later wrote to Dale Bumpers, chairman of the Subcommittee on the Upper Atmosphere of the Senate Committee on Aeronautical and Space Sciences, to refute the industry's claims. "I was speaking loosely to attempt to convey the idea that all predictions like this are subject to constant revision. In fact, I went on to state that the predicted ozone losses might change again. . . ."

In his own testimony before the subcommittee, Sherry Rowland explained it this way: "If I give you $5.00 and you give me $5.00 and I give you $5.00 and you give me $5.00, then I can say I gave you $10. But you don't have any more money. If the calculated ozone depletion goes up by a factor of two and down by a factor of two and up by a factor of two and down by a factor of two, if you quote the two down factors, you can say it went down by a factor of four, but you have left out the two up factors. I think that is what they have done."

Reporting on its own attempts to clarify Du Pont's use of the 300 per cent figure, *Rolling Stone* recounted this saga: "When Du Pont representatives were pressed for the precise meaning of that figure (which appears prominently under the heading 'Fact' in their October newspaper ads), it triggered a chain of baffling telephone exchanges wherein the figure was, depending on the statistical methods employed, revised to either 50 per cent or 75 per cent and by the end of which one representative admitted that 'we've had great arguments around here . . . you can't imagine how confusing it got.' "[3]

In the end, the National Academy panel concluded that the uncertainties in the chemical-reaction rates added a factor of about 2.5 to the total uncertainties of the final predictions of ozone depletion.

The most important question of course, is how good are the models—how well do they represent the real atmosphere. This is where atmospheric measurements came in.[4] Through the use of balloons and aircraft, scientists were able to sample the air in the stratosphere.

[3] The inconsistencies here are not quite as large as they seem. For example, if a number is reduced from 3 to 1, it has been reduced by 66 per cent. Looked at another way, 3 can be considered to be 300 per cent of 1.
[4] It was ironic, in view of the importance of these measurements in the fluorocarbon debate, that one set of experiments was seized on by congressmen who were critical of the methods of supplying funds for basic research. Congressman Robert Michel of Illinois, who was complaining about the obscure titles of the research projects that received grants, got up in the House of Representatives in April 1975 to voice these complaints. He had evidently heard little about the ozone controversy, because he obviously found the words "stratosphere," "chlorine pollutants," "fluorocarbons," and "ozone" unfamiliar. His comments: "I was equally excited to learn that David G. Murcray of the University of Denver will get $40,100 of the taxpayers' money to conduct a 'Measurement of the Stratospheric Distribution of the Fluorocarbons and Other Constituents of Interest in the Effect of Chlorine Pollutants in the Ozone Layer.' At least in that title there was one word I understood: constituents. And so the thought occurs to me, if one of my constituents should ask me my feelings about Dr. Murcray's project . . . what in the world could I possibly say?"

Stratospheric measurements served two important purposes. Some measurements serve as input data for the computer models—data that help the modeler to construct his computer simulation of the way the atmosphere is right now.

Other atmospheric measurements help scientists to check whether their models are right. The best test for a model is to determine how well it accounts for the distribution of chemicals in the existing atmosphere. There are several kinds of checks: Are certain chemicals actually present in the atmosphere in the amounts predicted by the model? Do their amounts change with height the way the models indicate they would? Do they change from day to night, as the sun turns off and on, in the way the models say they should? If the measurements made in the atmosphere are consistent with what the model predicted, this means you can put greater faith in everything else the model predicts—including things you can't check independently, such as what fluorocarbons will do to the ozone layer ten or twenty years from now.

Measurements proved to be crucial in demonstrating every step of the Rowland/Molina theory. There were three links in the chain of reasoning. The first was that the fluorocarbons actually got into the stratosphere, that they were not removed by processes in the lower atmosphere (more on this later). The second was that fluorocarbons, when they did reach the stratosphere, were split apart by the sun's energy to produce chlorine atoms. The third was that the chlorine atoms catalytically destroyed ozone.

The industry challenged the theory every step of the way. They said there was no proof that fluorocarbons even got into the stratosphere, no proof that they split apart to produce chlorine, no proof that, even if they did so, the chlorine was destroying ozone.

This contention systematically began to fall apart during the last part of 1975 and the early part of 1976. In December, Dave Murcray of the University of Denver reported that concentrations of fluorocarbons had more than doubled in the stratosphere over New Mexico between 1968 and 1975. Then, in the spring of 1976, there was another blow to industry's position: Researchers at the National Oceanic and Atmospheric Administration and at the National Center for Atmospheric Research, both in Boulder, Colorado, reported the results of separate experiments in which, for the first time, measurements of fluorocarbons had been made in that region of the stratosphere where the Rowland/Molina theory says they would be broken up by sunlight.[5] This was proof that the fluorocarbons were actually getting up to the predicted altitude, but it was more than that. The measurements also provided very strong evidence for the second link in the chain of reasoning—that the fluorocarbons were split apart by the sun's ultraviolet radiation— because the researchers had found the amounts of fluorocarbons dropping off with altitude in exactly the way the theory had predicted they would.

John Gille of NCAR said the data "strongly support the ozone-depletion theory and drastically reduce any room to deny the problem."

On the other hand, Ray McCarthy is quoted in *Environmental Science and Technology* as saying that the measurements "do not support predictions of ozone depletion. All they do is confirm that *some* fluorocarbons rise into the upper stratosphere." (Emphasis added.) As we shall see shortly, the industry's next step was to question what proportion of the fluorocarbons released at the ground actually made it up into the stratosphere.

What did the measurements actually show? In addition to proving that the fluorocarbons did indeed get into the stratosphere, the measurements also demonstrated that the chemicals were being broken up by ultraviolet light. This was indicated by the fact that the two different kinds of fluorocarbons—F-11 and F-12—showed a different pattern of breakup—a pattern consistent with the way these two broke up with ultraviolet as measured by Rowland and Molina in the laboratory.

The NOAA and NCAR measurements did not, in themselves, prove that the fluorocarbons were actually giving off chlorine atoms in the stratosphere. The final piece of the puzzle was added when laboratory experiments at the National Bureau of Standards

---

[5] Earlier measurements of fluorocarbons in lower regions of the stratosphere or near the upper part of the troposphere were made by a variety of researchers, including Jim Lovelock, Philip Krey (Energy Research and Development Administration), Peter Wilkniss (Naval Research Laboratory), John Swinnerton (NRL), and Norman Hester and O. C. Taylor (University of California, at Riverside).

(NBS), by Pierre Ausloos and Richard Rebbert, provided convincing evidence that fluorocarbons broken up by ultraviolet light do give off chlorine atoms. These data, together with the NOAA and NCAR measurements, provide strong evidence that fluorocarbons do in fact give off chlorine atoms in the stratosphere.

The NBS findings were not startling news—few in the scientific community had expected otherwise—but it was a new piece of evidence in support of the Rowland/Molina theory and it was so interpreted by the media. Du Pont was quick off the mark with a "clarification" protesting the interpretation of the results as support for the theory and pointing out the "real significance" of the work. "Actually, the study raises new questions about the completeness of the computer study used to predict ozone depletion." Du Pont contended that the NBS researchers had studied the role played by cold stratospheric temperatures in the breakup of fluorocarbon molecules, and this factor had not been included in previous computer models.

It was true that this temperature dependence had not been included in the models; as we have seen, models are approximations, and many things are necessarily left out. One of the recurrent themes in industry's campaign against the Rowland/Molina theory was to pick up any new thing that had not been included in the models and use this to cast doubt on all the predictions of the models, whether the omitted data had an important effect or not. The temperature dependence, now included in the models, did not have a big effect. It changed the ozone recovery time from ninety years to one hundred years in the case of one fluorocarbon (but did not change the other) and had no effect at all on the predictions of the absolute amount of ozone depletion caused by the chemicals.

The next step in industry's argument was that even if fluorocarbons broke up in the stratosphere, there was no proof that the fragments would behave in the way the models predicted. In particular, they challenged the ultimate end result of the theory—whether a repeating chain reaction between chlorine and ozone would occur. The Du Pont brochure described this as the "most speculative and most critical aspect of the theory" and also "the most difficult area to research."

The data to support this last part of the theory came in bit by bit.

The first piece of evidence was the measurement of HCl, one of the chemical products predicted by the models. The fact that HCl increased with height tended to support the picture of chlorine atoms being given off in the stratosphere by fluorocarbons, rather than percolating up from volcanic eruptions or sea spray. But the most direct proof would be the detection of chlorine atoms or chlorine oxide (ClO), the two vital links in the catalytic chain that destroys ozone. The only reaction scientists know of that produces ClO is the reaction, between chlorine atoms and ozone; the detection of ClO in the stratosphere would mean that chlorine had to have reacted with ozone. Thus, to most scientists working on the fluorocarbon problem, ClO was unquestionably the "smoking gun" they were looking for. Cicerone said that detection of ClO in the amounts predicted would "be enough" to prove the theory.

Industry, however, seemed to regard ClO measurements as a necessary, but not sufficient, condition for proving the theory. While emphasizing that ClO measurements were, in Ray McCarthy's word, "crucial," they deftly avoided being pinned down to an admission that such measurements would ultimately satisfy them of the guilt of fluorocarbons. (As we shall see, they didn't.)

Because chlorine atoms and ClO were present in the stratosphere in such tiny amounts, they were extremely difficult to measure. The Academy report was completed before there was a really good measurement of ClO; but, as it turned out, the regulatory agencies decided they had enough evidence without the smoking gun, for they started the regulatory process in late 1976 without the ClO data.

By late 1976, Jim Anderson of the University of Michigan had detected ClO in the stratosphere. This proved that chlorine does react with ozone. Anderson also found that the ratio of chlorine atoms to ClO was very close to that predicted by the computer models, which showed that the catalytic chain was indeed occurring. Some of his measurements, however, present problems. The most serious of these is that he finds too much ClO—more than can be explained from any known sources of chlorine, natural or manmade. The study of this problem continues.

But even before ClO measurements were obtained, the industry had already shifted its emphasis to the next stage of the argument.

Although detection of ClO would show that chlorine atoms reacted with ozone, it would not prove that the catalytic chain was as long as the theory predicted—that is, it would not show how many ozone molecules were destroyed by each ClO molecule. To do that would require a body count of all the chlorine atoms in the stratosphere to compare with the measured amount of ClO. Measuring total chlorine is, however, even more difficult than measuring ClO and would take a few years to do. Nevertheless, by early 1977, the Manufacturing Chemists Association, which administers the industry funds used for research, was focusing very strongly on the need for experiments to measure total chlorine.

It is important to examine the usefulness of the atmospheric measurements in testing the Rowland/Molina theory and the computer models on which projections of ozone depletion were based. The atmosphere *is* a complex place, as industry so often pointed out, and the measurements, however inadequate and difficult to obtain, were absolutely necessary.

(Not everyone thought so. Early in the fluorocarbon debate, Hal Johnston, in a letter to the National Academy panel, took an extreme position that such measurements would be "superfluous" to proving the basic elements of the Rowland/Molina theory. In part, he was reacting here to what he regarded as slurs on the CIAP program by the fluorocarbon industry. One industry-sponsored study had shown that one reaction worked in the opposite direction in the case of chlorine than it did in the case of NO_x, and some in the industry tried to dismiss the relevance of the entire CIAP program on these grounds. Johnston went to the other extreme. He told *Science News* that the CIAP study represented a large body of data "directly applicable to the [fluorocarbon] problem." There were "exact parallels" between the NO_x and the chlorine problems and "we don't have to redo all the basic work." The chlorine model was "virtually proven by analogy.")

If the measurements were important, however, it was always necessary to recognize their limitations. There *were* uncertainties, sometimes quite large uncertainties, and both Sherry Rowland and industry scientists were guilty on occasion of overinterpreting results at the limit of the uncertainties in the light most favorable to their own cause.

What can be said, then, about the message contained in the atmospheric measurements? In general, one can conclude that the measurements and the computer models do agree, taking into account the admittedly large uncertainties in the models and the fact that the measurements have been made in only a few locations at specific times, while most of the models deal with globally averaged conditions.

If that seems less than direct confirmation of the accuracy of the models, there was one event in 1972 that comes closer than anything else to such direct confirmation. In August of that year, an enormous flare occurred on the sun—the largest ever recorded—and it sent a large quantity of highly energetic protons streaming toward earth. In 1975, on the basis of his computer model, Paul Crutzen predicted that the top of the ozone layer should have been depleted 15 to 20 per cent over the North Pole, due to the production of NO_x from the interactions of the solar protons with the earth's atmosphere. (The effect would occur primarily over the poles because protons are focused there by the earth's magnetic field.)

Two NASA scientists, Donald Heath and Arlin Kreuger, then retrieved previously unanalyzed data from the Nimbus 4 weather satellite, which had been monitoring ozone levels during the 1972 flare. The data gave dramatic confirmation of Crutzen's prediction—a 16 per cent drop in ozone over the North Pole.

By mid-1976, there was no longer any real doubt that fluorocarbons were getting into the stratosphere: The NOAA and NCAR stratospheric measurements had clinched that. But industry spokesmen began to question *how much* of the total amount of fluorocarbons that had been released at ground level actually got up into the stratosphere. They steadfastly argued that there were removal processes—known as "sinks"—in the lower part of the atmosphere, the troposphere.

The search for tropospheric sinks occupied much of everybody's time, particularly that of the National Academy panel. No one knew of any such sinks. As we saw in Chapter One, Mario Molina looked for sinks and could not find any; several other researchers had also looked without success. Art Schmeltekopf of NOAA said that the stratospheric measurements made by his team demonstrated that fluorocarbons get into the stratosphere "with no large losses in the lower atmosphere." He suggested "it is unlikely we will find major tropospheric sinks."

This was true if by "major" he meant a large or efficient sink. But a removal process in the lower atmosphere would not have to be very efficient to be significant. There are two reasons for this. First, the movement of fluorocarbons up into the stratosphere, and their decomposition by sunlight, is a very slow process. If you were to stop all releases of fluorocarbons today, it would take about a hundred years for the amount of fluorocarbons in the atmosphere to drop to half its present value as a result of its decomposition by ultraviolet radiation in the stratosphere.

Since fluorocarbons stay so long in the troposphere, a tropospheric removal process, even if it works very slowly and inefficiently, will still have plenty of time to get the fluorocarbons before they escape into the stratosphere. Let us suppose that all releases of fluorocarbons into the lower atmosphere have stopped. Consider now a tropospheric process that reduces the fluorocarbons by half of the existing amount in a hundred years; this, as we have seen, is the same as the rate at which the stratospheric, or ozone-destruction, process is removing fluorocarbons. Thus these two processes would compete for the fluorocarbons on equal terms. The stratospheric sink would be only half as effective in such circumstances, and thus the amount of ozone destruction would be cut in half. Now, a tropospheric removal process that reduces fluorocarbons by half in a hundred years is equal to one that reduces them by .5 per cent in one year. A tropospheric sink of .5 per cent per year could have been overlooked by Rowland and Molina in their initial examination of various sinks.

There are two ways of looking for sinks in the troposphere. The best way would be to use the inventory, or budget, technique. Take the total amount of fluorocarbons that has been released to date, subtract the amount that the theory says has been used up in destroying ozone, and that should equal the amount still in the atmosphere *if* there is no other removal process. If the amount measured in the atmosphere is less than that, there must be another removal process.

But there was a problem with this approach. There were too many uncertainties—both in the figures for worldwide production and release of fluorocarbons and in the techniques for measuring the chemicals in the atmosphere.

The fluorocarbon industry came up with excellent statistics for the worldwide production of the chemicals; these figures were accurate to within 5 per cent. But total production was not the only factor of importance; it was also necessary to determine the rate at which the chemicals were actually released into the atmosphere. Obviously, they were released fairly quickly from spray cans, but there would be a delay time of many years for the release of fluorocarbons from hermetically sealed refrigerators. The industry was also able to come up with good figures for these delay times. The upshot was that the total uncertainty in our "budget" calculation attributable to the production and release of fluorocarbons was about 10 per cent.

The uncertainties associated with techniques for measuring fluorocarbons in the atmosphere were considerably higher. There were two difficulties here: first, there were inaccuracies in each individual measurement; second, since it was difficult to take measurements everywhere on the globe, it was hard to obtain a figure for the total amount of fluorocarbons in the atmosphere. Individual scientists would claim an accuracy of about 10 per cent in their measurements, but what this usually meant is that any two sets of measurements taken with the same instrument would agree with each other to within 10 per cent. It did not necessarily mean that the measurements were within 10 per cent of the amount of fluorocarbons actually in the atmosphere. (A temperature analogy will help us here again. Say you had a thermometer which, unknown to you, had an incorrect temperature scale. You take several measurements of body temperature and find that they all agree with each other to within 10 per cent—except that the absolute reading you are getting is around 70 degrees F., when normal body temperature is really 98.6 degrees. In other words, the measurements can be consistent but wrong, due to inherent errors in the instrument.) [6]

[6] It should, however, be noted that making these fluorocarbon measurements was no trivial matter. Measuring a few parts per trillion—once described as finding a single drop in a thousand tankcars—required measuring techniques never before developed.

A workshop was held in Boulder in March 1976 in an effort to resolve this problem. Each of the researchers who had been making atmospheric measurements of fluorocarbons brought an air sample. These samples were analyzed by two different groups in Boulder, and the results were compared not only to each other, but also to the results obtained by the researcher who had brought that particular air sample. (This meeting was called a "plow-off" by Art Schmeltekopf of the NOAA labs in Boulder—a phrase borrowed from his Texas farming background in which manufacturers of agricultural equipment would compete in plowing fields to demonstrate their wares to the farmers.)

To the dismay of all concerned, the plow-off succeeded only in demonstrating that the different researchers really did not agree with one another very well at all—in fact, their figures for the amount of fluorocarbons actually in the atmosphere differed by a whopping 40 per cent. Since then, the figures have been refined, but the total uncertainties in the "budget" calculation is still about 25 per cent.

What this meant was that one could either argue that there were no removal processes in the lower atmosphere or, equally well, that there were appreciable removal processes. Thus the atmospheric measurements—important in refining and testing the computer models—proved not to be very useful in the search for tropospheric sinks. Since many conclusions could be drawn from them, no definitive conclusion could be drawn from them.

This did not, of course, prevent both sides in the debate from drawing conclusions. At one point, industry spokesmen were arguing for the possibility that 80 per cent of the fluorocarbons could be removed in the lower atmosphere before reaching the stratosphere, while Sherry Rowland was contending that the data gave no evidence for a tropospheric sink of any kind—in fact, he argued that the measurements indicated there was more of one kind of fluorocarbon in the atmosphere than the industry had claimed it had made, concluded that the production estimates released by the industry were therefore too low, and jokingly suggested that perhaps industry had been cheating on its reports of fluorocarbon production to the Tariff Commission!

* For its part, the National Academy panel decided that this whole

approach was utterly useless. Instead, the panel members devoted
their efforts to a more painstaking and tiresome approach. They
began methodically to examine each and every removal process that
they or anyone else could think of. The suggestions ranged all the
way from freezing fluorocarbons out on the polar ice caps[7] to having them eaten by bugs in the oceans to having them destroyed by
ions in the atmosphere. There were a number of rather wild suggestions from researchers who had never previously done any work in
stratospheric chemistry. In their press releases, the industry
suggested a large number of highly speculative and implausible
sinks, all of which proved groundless. (Here, more than anywhere
else, the industry press releases demonstrated an ignorance of science. One Du Pont ad, for example, suggested that "ion molecules"
might react with fluorocarbons to remove them from the atmosphere. There are no such things as ion molecules. The proposed sink
referred to actually dealt with reactions *between* ions and molecules. Industry scientists persistently disclaim any responsibility for
ads such as these, saying they were the work of the marketing and
advertising departments. But they cannot get off scot-free on this.
One is left to wonder why they were not consulted for scientific content or, after such errors appeared in print, they did not take it
upon themselves to see to it that the errors were not repeated.)

But the industry never gave up hoping for a tropospheric sink.
There was a certain irony in this—an irony having to do with the
industry's own early research into the environmental impact of
fluorocarbons. In February 1975, a Du Pont *Management Bulletin*
noted that "last year, reports on research begun by the fluorocarbon
industry in 1971 showed that the fluorocarbons had no adverse
effect on plant life, did not contribute to the formation of smog,
and were not decomposed by the usual chemical reactions near the
earth.

"Also in 1971, it was reported that the compounds are not
scavenged from the atmosphere by physical and chemical
processes."

[7] Reinhold Rasmussen, a Washington State University researcher, found
snow and ice samples collected at the South Pole to be enriched in
fluorocarbons. He suggested that the chemicals were removed from the
troposphere by ice and snow, but this has not been confirmed by observations made in other locations.

Although the industry representatives never couched it in such terms, what was implicit in this statement was that their own search for tropospheric sinks had also been singularly unproductive. Of course, at the time, the industry was not looking for tropospheric sinks as such and, as a matter of fact, they were rather hoping they wouldn't find any. If they had, this could have branded fluorocarbons as pollutants. So, before 1974, the fact that they couldn't find any removal processes in the lower atmosphere was "good news." After 1974, it was "bad news."

The industry's claim that "something might turn up" in the way of a tropospheric sink was irrefutable. The obvious failing of the tactic of searching for individual sinks was that you could never be absolutely sure you hadn't overlooked something. An example of this occurred after the NAS report had been completed, when scientists at the National Bureau of Standards (NBS) suggested that sandstorms might remove fluorocarbons in the lower atmosphere. This suggestion resulted from observations made by Jim Lovelock and experiments conducted by the NBS. On one of his ship cruises, Lovelock noticed that the concentrations of carbon tetrachloride dropped off in the atmosphere near the Sahara. Subsequent laboratory tests with sand at the NBS revealed the fluorocarbons adsorbed onto the surface of sand particles could, unlike free-floating fluorocarbon molecules, be broken up by the wavelengths of sunlight that reach the earth's surface. The list of possible tropospheric sinks had now run the gamut from the freezing poles to the baking desert. It appears, however, that sandstorms will not be an important "out" for fluorocarbons. But this incident does demonstrate that there is always the possibility of an unknown sink lurking in the background.

A proportion of the scientific studies that were conducted during the crucial 1974 to 1976 period were funded by the fluorocarbon industry. In late 1974, the industry announced that $3 million to $5 million would be made available over the next few years to university researchers working in the field. This money was contributed by twenty fluorocarbon manufacturers in nine countries, including the United States, Canada, Australia, Japan, and several European countries. Early in 1975, the Manufacturing Chemists Association (MCA), which was administering the fund, announced that ten projects had been accepted for that year. Three of them continued research funded by the industry in previous years; Jim Lovelock's program of measuring fluorocarbons in the lower atmosphere was among them. (Although he had to bankroll his initial research himself, Lovelock later received industry funds to continue his work.)

The projects announced by the MCA in early 1975 were strongly oriented toward measuring fluorocarbons in the upper atmosphere and toward the more difficult task of finding a way to measure chlorine atoms and ClO in the stratosphere. MCA also funded projects to measure chemical-reaction rates in the laboratory. By the end of 1976, nearly fifty separate projects had been funded. Throughout the fluorocarbon debate, the MCA handled itself in a very professional way in administering its research program and in its use of the data obtained. It wisely chose to fund primarily university researchers who were already working in the relevant fields—thereby successfully avoiding any charge that the industry was trying to "stack the decks." The MCA also used reputable scientists as consultants and referees in evaluating the research proposals. Finally, all researchers were permitted to publish their results in the scientific literature and to interpret them without industry censorship no matter where the chips fell. (It is doubtful that the researchers would have accepted any other conditions.)

Some in the industry argued that regulatory action should await the results of the industry's program. "Until the first results of this industry research become available," said Ray McCarthy, "the available facts do not rank as proof that the fluorocarbons will lead to ozone depletion or even that the claimed chain reaction could occur at all."

As far as Sherry Rowland was concerned, the industry had already had its chance. They had been making fluorocarbons for more than forty years, and they had neglected over that time to build up any expertise in stratospheric chemistry. "I don't think people should be rewarded with a grace period for not doing a thorough job of finding out the consequences [of their activities]," he is quoted as saying in the *Orange County Illustrated* magazine. "I consider ignorance a poor excuse. . . ."

The confrontations that had been occurring between scientists and the fluorocarbon and aerosol industry in the political and public-relations spheres had their counterparts in the scientific sphere. Every new atmospheric measurement, every new computer calculation, every new laboratory test, every new candidate for a tropospheric sink was immediately trotted out for minute inspection, divergent interpretation, and heated debate. Even the scientific meetings became highly emotional at times. An American Chemical Society meeting in Philadelphia in April 1975 was described by industry representatives as a "cozy gathering of hate." An account of the meeting in the Philadelphia *Inquirer* described Du Pont's scientists—"some of whom appeared on the verge of losing their tempers"—as beleaguered and outgunned by their scientific opponents. Sherry Rowland was depicted in the aerosol trade press as engaging in "scientific sniper fire."

Not surprisingly, the industry was highly critical of what it regarded as crusading on the part of its scientific opponents. Vincent Carberry, head of advertising and public relations for Precision Valve, grumbled in *Harper's Weekly* that "just because a guy has a few Ph.D.'s after his name, I don't think that gives him any right to shoot off his mouth when he's not certain."

It was not only the aerosol trade press, however, that noticed the high PR profile of scientists. In December 1975, *Rolling Stone* magazine wrote of the "fairly intensive campaign" waged by the National Bureau of Standards to publicize research results by two of its researchers (Ausloos and Rebbert). "The research was interesting," *Rolling Stone* noted, ". . . but it was not earthshaking." The extent of the publicity just before a scientific meeting at which the results were discussed "puzzled many of the convention attendees and apparently thoroughly intimidated the researcher involved, who proceeded to make himself just about invisible during the course of the meeting."

Rolling Stone also noted that "the publicity efforts of . . . Rowland seem unusually vigorous for scientific circles, and Rowland himself is quick to admit that between his speaking tours he regularly spends up to two hours a day on the telephone to media. Rowland is rather media hip for an academic scientist; most institutions find it necessary to provide their researchers with professional publicity staffs."

Rowland was, of course, not the only scientist with a high public profile in the fluorocarbon debate. Virtually all researchers who had anything to do with it had some exposure to the press from time and time and, at the height of the controversy, several of them were spending a good proportion of their time fielding reporters' queries, not always ungrudgingly. Many of these scientists seemed to relish this activity and hate it in about equal measure; it stole precious time from the doing of research, and the frustrations of dealing with media excesses and errors were plentiful. But, on the other hand, their names were in lights, and that was not altogether an unpleasant experience. Most of the scientists co-operated with the press, or selected representatives thereof, with varying degrees of willingness. However, except for a few instances here and there, the scientists could not, in general, be accused of actively peddling their wares to the press; it was more a case of the press harassing them than the other way around. (McElroy says he got tired of it very quickly, especially after receiving calls from reporters at two in the morning.)

It was ironic that the industry chose to complain about the PR tactics of scientists when it conducted so much of its own scientific battle through ads, press releases, and other vehicles of "clarification" that proclaimed the scientists' announcements to be "misleading," "incomplete," "not proved," or "not news." It was interesting, however, to witness the impact that various scientific studies had on the tenor of the industry's attack and on the progression in their argumentation.

Consider again the three steps in the Rowland/Molina theory: (1) Fluorocarbons get into the stratosphere without significant losses to sinks in the lower atmosphere; (2) they decompose with ultraviolet radiation to produce chlorine atoms; (3) the chlorine atoms then catalytically destroy ozone.

In September 1975, Ted Cairns, director of Du Pont's Central Research and Development Department, issued the following statement: "No one has seriously questioned whether fluorocarbons would rise to the stratosphere, that they would be broken down by ultraviolet light, or that chlorine will react with ozone." (He did, however, question whether there was a repeating chain reaction, saying that each chlorine atom must destroy tens of thousands of ozone molecules for the theory to be "of any public importance.")

It is interesting, in light of Cairns' assertion, to examine the record of industry statements that preceded it.

EXAMPLE: Dorothy Smith, news manager of the American Chemical Society, states in a letter to science writers that in September 1974, "a public-relations official of a major [fluorocarbon] producer assured me that the theory was weak, there was no known mechanism for chlorofluorocarbons to reach the stratosphere, and none had been detected there."

EXAMPLE: *Chemical and Engineering News,* September 1974: "Materials do not diffuse uniformly throughout the atmosphere, [Ray McCarthy] says, and it has not yet been well established that fluorocarbons actually penetrate as far as the ozone layer. . . ."

EXAMPLE: New York *Times,* November 1974: "Spokesmen for the multimillion-dollar industry . . . have argued that there is no evidence for such upward movement of the gases or for their breakdown under stratospheric conditions."

EXAMPLE: *The Sciences* magazine, December 1974: "No measurements have been made of the presence, much less the breakdown, of the fluorocarbons at stratospheric levels—a point emphasized by Raymond L. McCarthy . . . at the ACS meeting."

EXAMPLE: Du Pont *Management Bulletin,* February 1975: " 'There is no concrete evidence to show that the ozone-depleting reaction with chlorine takes place,' Raymond L. McCarthy said in testimony before a House Commerce subcommittee."

EXAMPLE: *Food and Drug Packaging,* March 1975, an interview with H. R. Shepherd, president of Aerosol Techniques, Inc.: "The industry has a number of studies under way to answer some of the basic questions. For example: . . . Are fluorocarbons in the stratosphere? . . . Do they in fact break down and release a free chlorine radical? Does that chlorine radical react with ozone, or with other substances, or both?"

EXAMPLE: Hearings before the Senate Space Committee, September 1975, written statement of Roy Schuyler of Du Pont: "Perhaps of greatest importance, it is not known whether fluorocarbons *do* enter ozone-depleting chemical reactions in the stratosphere." (Emphasis his.)

EXAMPLE: *Aerosol Age*, December 1975, report of a speech by Ralph Engel, executive director of the Chemical Specialties Manufacturers Association: "We do know that fluorocarbons reach the upper atmosphere. . . . We do not know what occurs when fluorocarbons attain stratospheric altitude."

EXAMPLE: New York *Daily News*, January 1976, interview with Ray McCarthy: "There is very little experimental data to show whether the fluorocarbon dissociation takes place and whether the ozone-depleting reaction takes place."

Yet despite all this, a Du Pont brochure would later claim that "some elements of the theory have been generally accepted all along and confirmed by recent experimental work. These include the fact that some fluorocarbons will rise to the stratosphere and that they will be broken down by sunlight." (The brochure, of course, went on to question various other aspects of the theory, the models, and the experimental data.)

Almost without exception, the studies of 1975 tended to undermine the industry's position, although, as we have seen, there was always enough ambiguity and uncertainty to leave room for considerable argument. And there continued to be plenty of that. Some truly amazing claims were made by Jack Dickinson, head of COAS and vice president for consumer affairs for Gillette, who had these views to offer the Bumpers Senate Subcommittee on the Upper Atmosphere in September 1975:

"Virtually all of the recent data and findings have done one of three things. They have either added atmospheric knowledge without making any real contribution to confirmation or refutation of the hypothesis, or they have caused the predicted reactions to be revised downward substantially . . . or they are inconsistent with the hypothesis." Later, in early 1976, Dickinson suggested the hypothesis had not been subjected to normal scientific experimentation and observation and that the findings had so far been of little significance.

This was so much whistling in the dark. Unlike Dickinson, many in the industry were, by the end of 1975, being forced by the accumulating scientific evidence to shift their emphasis somewhat. No longer could they attack the Rowland/Molina hypothesis as "just a theory" unsupported by experimental evidence. Instead, they began increasingly to suggest that the magnitude of the effect would be small—that even if the theory were demonstrated to be qualitatively correct (and they did not concede that it had been), it would nevertheless prove to be quantitatively unimportant.

This shift had an important bearing on the industry's continuing campaign to delay regulations for two or three years. Several scientists had not been as adamant as Sherry Rowland about the need for immediate regulation and stated that waiting a couple of years would not be catastrophic. (The industry happily cited these comments—usually summarized with the statement that "most scientists agree there is time to study"—while carefully ignoring the growing consensus among scientists that the Rowland/Molina hypothesis was qualitatively correct.) Abandoning the position that there was no proof that ozone depletion would occur at all, the industry argued that the *amount* of depletion that would occur during a three-year delay would be "an insignificant and undetectable one half of 1 per cent." They got this figure from a computer study by Paul Crutzen (who was less than pleased with the manner in which the industry used the data). Crutzen had calculated that if fluorocarbon releases were curtailed in 1975, the probable reduction in ozone by 1990 would be 1.2 per cent. If releases were stopped instead in 1978, the probable reduction would be 1.7 per cent. The "one half of 1 per cent" quoted in congressional testimony by Du Pont's Roy Schuyler represents the difference between these two figures and is the "incremental ozone depletion" that might be expected as a result of a three-year delay in regulations. Du Pont was fond of arguing that the risk of this delay "is about the same risk taken when a person moves his home thirty-five miles South." We have already seen why this is a specious argument, but there is one other point worth raising here: That one half of 1 per cent additional loss of ozone does not occur just in one year, or even just in the three years of the delay; it occurs *every year* for the number of years that the ozone depletion continues, which could be as long as a century. Consider this: Say that if you banned fluorocarbons in

1978, you got one half of 1 per cent less ozone in 1990 than you would have had if you had banned them in 1975. You would also get one half of 1 per cent less in 1991, 1992, and so on for perhaps a hundred years. Suppose that that ozone depletion of one half of 1 per cent caused six thousand additional cases of skin cancer. *Each year,* you would have six thousand *more* cases than you would otherwise have had if you had banned the chemicals three years earlier. If the depletion continues for a hundred years, the "risk" associated with that three-year delay amounts to six hundred thousand additional cases of skin cancer.

It is, of course, a value judgment, not a scientific judgment, whether that risk is worth taking for the sake of three additional years of using fluorocarbon-containing spray cans.

By the end of 1975, the fluorocarbon debate seemed to be sorting itself out. Though the arguments were continuing, the National Academy panel, at least, felt pleased with the progress it was making. They had looked at all the new data and felt that they had come to grips with them. They had pondered the uncertainties in the models, in the transport of chemicals, in the reaction rates. They had searched diligently for sinks. The remaining uncertainties it seemed to them, were small enough that a report could be prepared and it could be taken as a basis for regulatory action. They began to write. A draft of the report went out to the reviewers, and the word soon came back that the reaction of these scientific referees was very favorable.

Then Sherry Rowland started talking about chlorine nitrate.

CHAPTER TEN

"Ten Plus Ten Equal Twelve"

New evidence has brought the predicted impact of fluorocarbon compounds in the earth's ozone nearly to zero. . . .

—Aerosol industry press release, May 12, 1976.

It really doesn't change the picture that much. In some ways, it makes it worse—more complicated, but no less serious.

—MARIO MOLINA, in *The Observer*, May 16, 1976.

The debate over chlorine nitrate, which occurred in the spring of 1976, once again threw the fluorocarbon controversy into a state of utter confusion. It temporarily derailed the National Academy panel. Along with other new evidence, it caused many in the aerosol industry to climb very far out on a limb (which was promptly sawed off). It caused yet more dissension within the scientific community. And it very probably caused an erosion in the public credibility of scientists—for it generated a number of news stories that made it appear as though scientists really didn't know what they were talking about. A climate of mild hysteria prevailed for more than a month—all because of a chemical that was a little hardier than anyone thought it was.

Chlorine nitrate is a chemical formed by the reaction of chlorine oxide (ClO) with nitrogen dioxide (NO_2). The former is a vital

link in the chlorine chain that destroys ozone; the latter is a vital
link in the nitrogen chain that accomplishes the same task. The reaction between them renders both inactive with respect to ozone;
chlorine nitrate was thus described by one scientist as a "safe holding tank." It is immediately obvious that the chlorine nitrate reaction might well reduce the threat to the ozone layer, not only from
chlorine sources such as spray cans, but also from NO_x sources such
as SSTs (more on this later).

Of course, scientists knew about this reaction before the spring of
1976, but the Academy panel had dismissed it as unimportant because it was generally believed that the chlorine nitrate formed in
the atmosphere would be very short-lived, that it would be rapidly
broken up by sunlight. Sherry Rowland himself informed the panel
in the summer of 1975 that chlorine nitrate was very unstable and
would break up in a matter of minutes.

Rowland based this statement on measurements that had been
done by German scientists in the 1950s, but he was sufficiently un-
easy about the measurements that he decided to double-check. By
the end of 1975, he and his colleagues, Mario Molina and John
Spencer, had concluded that chlorine nitrate would be much more
stable in the atmosphere than anyone thought. Rowland calculated
that measurable quantities of chlorine nitrate would be formed in
the mid-stratosphere between twenty and thirty kilometers. Higher
up, chlorine nitrate would have a shorter lifetime because of the
increasing solar radiation. Rowland concluded that his estimates
of ozone reduction would have to be revised downward by perhaps
20 to 30 per cent because of chlorine nitrate.

To his credit, he was very forthright in informing the scientific
community about his new finding, even though it reduced the
seriousness of the fluorocarbon problem and therefore tended to undercut his political position on the need for an immediate ban. In
fact, it was fortunate, in light of the subsequent controversy, that it
was Rowland himself who pursued and revealed the chlorine nitrate problem.

In revising his own ozone-depletion estimates by 20 to 30 per
cent, Rowland had accounted only for the fact that chlorine nitrate
would take out chlorine; he inexplicably failed to consider that it
would also remove NO_x, the other ozone-eater. With both chlorine

and NO_x out of the way, the impact on the ozone layer might well be substantially reduced. In fact, Steve Wofsy of Harvard suggested that if you added enough NO_x to reduce ozone, say, 10 per cent, and enough chlorine to reduce it 10 per cent, you wouldn't get a 20 per cent reduction, because of interactions between the compounds. As Sherry Rowland put it, it appeared to be a case of "10 plus 10 equal 12."

The chlorine nitrate problem immediately generated a flurry of activity within the scientific community. In particular, the push was on to refine the rate constants (speeds) of the chemical reactions involving chlorine nitrate and to plug these new data into the computer models.

As was typically the case, the situation rapidly became more confused as soon as the computer modelers got their hands on the new data. The whole community was stunned when the earliest model calculations indicated that the introduction of the chlorine nitrate factor might actually result in net *increase* in ozone. For a brief period there in February 1976, it appeared that the whole problem had suddenly vanished! This led to the bizarre and only half-joking speculation that it would be O.K. to use spray cans as long as you had SSTs flying, and vice versa. Since one added chlorine to the atmosphere and the other added NO_x, they would neutralize each other when the two compounds formed chlorine nitrate.

To make matters worse, the chlorine nitrate problem was not the only new and startling development on the scene. There was another set of data that appeared to be in conflict with the fluorocarbon/ozone theory. Al Lazrus, a researcher at the National Center for Atmospheric Research, had been measuring hydrogen chloride (HCl) in the stratosphere. Like chlorine nitrate, HCl is also a "safe holding tank" (the only one that had been included in the computer models prior to chlorine nitrate). HCl does not react with ozone itself, but it can be converted to the forms of chlorine that will by chemical reactions in the atmosphere. Thus the impact on ozone depends on how much of the chlorine is present in these reactive forms (chlorine atoms and chlorine oxide or ClO) compared to the amount locked up in the nonreactive form, HCl.

In early 1976, Lazrus circulated a preprint of a scientific paper among atmospheric scientists reporting that the amount of HCl in the stratosphere dropped off much more abruptly with height than predicted. This needed some explaining: Where was the chlorine going? Lazrus (and Sherry Rowland) suggested that the chlorine from HCl might be transferred by chemical reactions into that other newly discovered safe holding tank, chlorine nitrate—a speculation that was not convincing to others and ultimately proved unfounded.

These developments understandably caused considerable consternation among the members of the National Academy panel, who could feel the April deadline breathing down their necks. Despite the fact that the panel report had been virtually complete and that the draft had received rave reviews, they knew they could not release it without considering the new data. The combined confusions were large enough that they realized they could not meet their April deadline.

Almost immediately, this decision started causing problems, for the Academy was under increasing pressure to get the report out. Most of this pressure was coming from government departments and agencies whose activities and plans were being held up in anticipation of the Academy report. These agencies were irritated, not only by the delay but also by the request to bankroll the inevitable costs entailed by the delay. The Academy had originally received $159,000 to do the study—$59,000 from NASA, $50,000 from the National Science Foundation, and $25,000 each from the National Oceanic and Atmospheric Administration and the EPA.

According to one knowledgeable Academy source, "it was very difficult to get the funding that allowed us to complete that study." He said the Academy was told, in effect, "You're just diddling around. There's always going to be new information. You scientists are never sure; you're always saying, 'Wait till we look at this and wait till we look at that.'" Some of the agencies "would have been just as happy if that Academy hadn't done the study." Many had gone on the public record, through IMOS, saying there was "serious cause for concern," and some had already established their own schedules for programs and regulation. The Academy was now screwing up that schedule.

In the dickering that followed, the Academy warned the agencies that it would abort the study rather than release an incomplete report that had not taken new information into account or received proper peer review from the scientific community. The agencies capitulated; the IMOS report, after all, had specified that action was to be taken if the Academy report confirmed the IMOS assessment. By the end of the study, the Academy had received a total of some $300,000.

Word of the delay in the Academy report—and of the new data that had caused it—got out quickly, and the media soon started hounding the scientists working on the problem. Some headlines suggested a complete turn-around in the ozone/fluorocarbon controversy: "Aerosol scare 'may be over,'" said the London *Observer*. Other headlines conveyed the sense of uncertainty gripping the scientific community: "Rumor and confusion follow ozone-theory revision," noted *Science News*. "Is the 'threat' of aerosols going pfffft?" asked the *National Observer*.

Initially, the Academy was tight-lipped about the new data, refusing to discuss them at all, although the scientific community and most interested reporters knew all about them.[1] Later the Academy would say only that it was studying the matter. Of course, this fired all kinds of speculation and rumor—and a good deal of second-guessing about the Academy report. "The panel is likely to have to revise its findings completely," the *National Observer* conjectured. "No one outside the NAS knows for sure, but it's possible that in a few weeks the panel will announce that fluorocarbons do not in fact pose an immediate threat to human life." (Of course, no one had ever suggested that they did pose "an immediate threat to human life.")

Behind closed doors, the Academy panel was becoming more and more exasperated with the chlorine nitrate situation. Reports were filtering back from the computer modelers that they were getting large and confusingly different results for the effects of chlorine nitrate, so the panel members applied their attention to trying to wrest a coherent picture from the modelers. The panel was determined to make the modelers resolve the inconsistencies—or at least

[1] At this crucial time, there was a "leak" to the media concerning the Academy study. See Chapter Eleven for an account of how the Academy handled this situation.

explain why they differed—even if they had to lock the modelers in a room with a computer to accomplish the task. This is in fact what the panel did; during a weekend at the end of April, Paul Crutzen, Ralph Cicerone, and Julius Chang (a member of the panel and a computer modeler from Lawrence Livermore Laboratory) settled down to some serious number-crunching with the computer at the National Center for Atmospheric Research in Boulder, Colorado. They stayed up till the small hours on Friday, Saturday, and Sunday and reported to the panel on Monday morning.

The Harvard team of Mike McElroy and Steve Wofsy were noticeable by their absence from this session. They had in fact refused to attend. McElroy had witnessed the eruption of the chlorine nitrate snafu from afar—he had been in Ireland at the time—and he later described the uproar as "the most embarrassing thing this field has seen in a long time."

McElroy was critical of Rowland's early presentation of the chlorine nitrate problem—by what he described as Rowland's attempt to "put it under the rug and say it was a nonproblem." The computer modelers, he said, "immediately saw it was potentially a very serious problem."

McElroy says that the computer models that existed at the time were inadequate to deal with the chlorine nitrate problem—that they essentially "meant nothing." He said that he and Wofsy decided to stay home and start building a better model, rather than attending the modelers' meeting in Boulder, which McElroy said "smacked of preparing a unified case for the press when the whole field was looking a little silly." Wofsy told one panel member that the Harvard scientists were afraid the panel would call a press conference after the session to tell the media what the modelers had concluded.

This was utterly ridiculous. The panel members had been avoiding the media like the plague since the chlorine nitrate issue had erupted the month before, and even if they had been personally disposed to call a press conference, they were under strict and explicit orders from the Academy to refer all media queries to the Academy's information office. The modelers' meeting was simply an internal workshop to help the panel understand the chlorine nitrate problem, and the panel members, as well as the other modelers, were furious with McElroy and Wofsy for not co-operating. Many felt that their reasons for not attending had less to do with a fear about media coverage than the fact that McElroy and Wofsy did not want to share the limelight.

In the end, the meeting in Boulder did not resolve all the inconsistencies, and the modelers' report on Monday morning succeeded only in depressing the members of the Academy panel. The panel's job was obviously not finished. The members could not go back to their own research—they would have to solve this problem while still under the gun from all the pressure groups clamoring for their report. Worse, the report would inevitably be much weaker than it had originally been; the uncertainties would be much larger. Fred Kaufman and Harold Schiff both grumbled that the report would probably have to be "wishy-washy."

As the panel went back to work, it became increasingly apparent that the delay in their report might last for many months, not the one or two that had originally seemed likely.

The industry was, of course, delighted by this turn of events. Although some in the industry, like the Manufacturing Chemists Association, did not jump the gun, others did not waste much time in trying to capitalize on the new developments. COAS was a case in point. On May 12, COAS called a press conference in New York to announce that the various bits of new data had brought the predicted impact of fluorocarbons on the ozone layer "nearly to zero." Outside the hotel where the conference was held, the New York Public Interest Research Group passed out "ban the can" leaflets; inside, Jim Lodge presided over the press conference, issuing a statement that claimed that Lazrus's HCl data raised doubts about the validity of the Rowland/Molina theory, and suggested that important chemistry had been left out of the calculations. This was to become the predominant theme of industry spokesmen and scientists, many of whom seemed to focus much more on the HCl data than on chlorine nitrate, initially thinking it a kind of miracle.

Lodge did not, however, ignore chlorine nitrate. At the press conference, he cited unpublished work by Douglas Davis, a University of Maryland researcher who was studying rate constants of important chemical reactions with industry funding. Lodge said that

Davis's measurements indicated that "the chlorine would remain trapped in the chlorine nitrate sufficiently long to dramatically reduce its impact on ozone." This is not precisely what Davis' data meant; his studies indicated that when chlorine nitrate broke up with ultraviolet radiation, it did not do it in the way that scientists had believed it would. Instead of releasing chlorine to do further damage, it apparently went into another compound stable enough to act as a permanent safe holding tank. However, these data were preliminary, and they shortly proved to be wrong.

At the press conference, Lodge claimed that adjusting the atmospheric models to account for chlorine nitrate "has reduced forecasts of ozone depletion by 60 to 90 per cent."

(Davis said in an interview at the time that he was unaware that his work was being cited in the COAS press release—he asked to have it read to him—and he said he did not necessarily subscribe to the 80 to 90 per cent figure. "That's one calculation," he said. "There are many calculations going on." Throughout this episode, Davis continued to treat his preliminary data with great caution when speaking to the media.)

At a meeting of the Chemical Specialties Manufacturers Association that spring, industry spokesmen, quoted in *Drug and Cosmetic Industry,* criticized "campus theoreticians" and announced that "there are big holes developing in the theory." Daniel Harnish, of Allied Chemical, said that "in effect, we have destroyed an essential premise of the theory, that all the necessary chemistry in the reaction was known."

For some reason, many industry spokesmen and scientists seemed to pin their attention and hopes mainly on Lazrus' HCl data. It was, ultimately, an unhappy choice from the scientific point of view. At the end of May, as they sat happily dangling their feet from the end of the limb they'd crawled out on, the branch was abruptly cut down. Al Lazrus distributed a "dear colleague" letter widely throughout the scientific community saying that the HCl data in his preprint paper had been wrong. A calibration error, combined with the unfortunate omission of a lab procedure for checking results, had led to the release of the incorrect information. The data no longer flatly contradicted the Rowland/Molina theory.

The media duly reported this "setback" for the industry. ("Du

Pont Blanches" read one trade magazine headline.) Donald Davis of *Drug and Cosmetic Industry,* pulling no punches as usual, gave industry its lumps: "After two trips out to the end of the limb [the first was during the volcano incident] there probably isn't time for another by COAS" before the NAS report.

Although some in industry—notably Du Pont scientist Peter Jesson and COAS adviser Jim Lodge—continued to insist that the Lazrus data still indicated that the theory's chemistry was all wrong, many industry spokesmen were reduced to the lame rejoinder that if Lazrus' data didn't demolish the fluorocarbon/ozone theory, something else (unnamed) would. The *Wall Street Journal* quoted Jim Lodge blustering that "the whole thing seems to resemble a structure held together with Tinker Toys, Scotch tape, and rubber bands."

Others, however, saw the handwriting on the wall. The *Journal* quoted a Dow Chemical Company scientist as saying: "Most of the research going on . . . is aimed at ironing out details. I don't think there's going to be any thunderbolt knocking out the theory."[2]

It was entirely typical of the fluorocarbon/ozone controversy that the Lazrus data did not have time to shake down scientifically before they were paraded out into the public limelight. In normal science, a slow and often laborious process of scientific review goes on; if it sometimes frustrates nonscientists, it at least has the salutary effect of preventing errors from becoming too public or too widespread. But the fluorocarbon controversy was, in many respects, not normal science; in the tense and fast-paced atmosphere that surrounded it, the process of scientific review and cogitation was at least greatly compressed and, on occasion, entirely circumvented by the media, by industry's public relations and sometimes by the scientists themselves. In the case of the Lazrus data, this may have been further compounded by the impending release of the Academy report; many in the industry were anxious to find something that would subvert the theory, and they were just waiting for a chance to say "I told you so."

[2] Although it produced some of the raw materials of which fluorocarbons were made, Dow was not itself a major producer of fluorocarbons. A Dow scientist later acknowledged that "we didn't have as big a stake to defend" as other companies like Du Pont did.

One who was more than ready was Robert Abplanalp of Precision Valve Corporation, who was evidently still smarting over the 50 per cent cut in business he had suffered in 1975. Precision Valve ran an ad titled "Who's out on a limb?" that featured a reproduction of a news story about Sherry Rowland with the headline "Key ozone-scare scientist changes tune after research."

The ad said that, at the beginning of the ozone controversy, the industry was alarmed, "but we could hardly believe that the aerosol industry, a concerned industry made up of conscientious suppliers and marketers who spared no expense to test their products' integrity and efficacy, could founder on the rock of a tenuous theory. . . . Suddenly we had the gut feeling we were the victims of overzealous environmentalists. . . . Well, it has taken all of two years and many millions of dollars of industry and government money to justify our gut feeling. Important new evidence does not support the ozone-depletion theory. . . . We are not home free yet, but we are on our way. . . . Keep the faith. Consumers prefer aerosols."

Industry was gambling, taking a calculated risk. The "new evidence" was decidedly preliminary, and there was no guarantee that it would ultimately confute the Rowland/Molina theory. But the confusion initially worked in the industry's favor; it sowed doubt in the public mind and provoked an exasperation with scientists who seemingly kept changing their minds. The gamble paid off handsomely; the New York press conference, for example, received wide coverage and garnered press statements like: "New doubts cloud the issue. . . ." and "Experts now see risks to ozone from fluorocarbons lessened." *Aerosol Age* gleefully ran a full-page pastiche of favorable headlines.

The anti-aerosol forces continued to insist that it was the industry that had blundered out on the limb—and in this they finally proved correct—but what they failed to realize is that the industry's tactics made sense strategically, if not scientifically. From a PR point of view, it really mattered very little that their use of the Lazrus data backfired. The headlines had been favorable and, as often happens, the retraction did not receive nearly as much coverage as the original story had. Ultimately, the public was left with the distinct impression that the fluorocarbon/ozone problem had been reduced "nearly to zero."

The seeming turnabout in the scientific situation, along with a number of other factors, made the spring of 1976 a very upbeat period for the aerosol industry. They were buoyed by the release of a report by the British Government that tended to minimize the seriousness of the problem and concluded that "there appears to be no need for precipitate action." The report did, however, recommend that the industry should intensify its search for alternative propellants and seek to minimize leakages of fluorocarbons from industrial equipment. (In general, British scientists had maintained their attitude that the Americans were overreacting, and the U.S. aerosol industry heartily approved of this "rational" and "level-headed" British approach.)

Things also seemed to be looking up economically. Shipments of aerosol cans for February and March were up 26 and 41 per cent, respectively, over the same months in 1975; the increase for the first quarter was 17 per cent over the previous year. The year 1975 had undeniably been a "miserable" one—aerosol production had dropped nearly 14 per cent from 1974—but, according to *Chemical Marketing Reporter,* some in the aerosol industry were, by mid-1976, beginning to "paint a glowing picture of a return to growth. . . ." *Aerosol Age* noted there was an "enormous feeling of optimism" at the CSMA's midyear meeting in Chicago; the predictions were that the 1976 production would nearly equal the record output of 1973.

"The initial scare is wearing off," one industry spokesman told *Chemical Marketing Reporter* in reference to the ozone controversy.

It is interesting to note that many in the industry were by this time discounting the ozone problem as the major cause of the aerosol slump. The economy received much of the blame; one industry representative showed a number of charts demonstrating the parallel between aerosol production and such things as gross national product, prime interest rate, unemployment figures, and personal expenditures.

As ever, Donald Davis of *Drug and Cosmetic Industry* was skeptical of all this optimism. He was unimpressed with the "elaborate statistical exercise to downplay the ozone-depletion assault on the aerosol's public image" and said that the dip in aerosol production was much greater than all the other factors. The erosion of 1975 was far too great to support the contention that the ozone controversy was not the major factor, he said. He was dubious that the optimistic forecasts for 1976 could be sustained without "statistical sleight of hand."

Interestingly enough, many in the aerosol industry attributed the failing fortunes of the aerosol industry primarily to the aggressive advertising of nonaerosols by marketers who had jumped ship. Rebounding economy notwithstanding, the aerosol industry continued to fret over the continuing encroachment of the nonaerosol. (In early 1976, nonaerosol deodorants had jumped 800 per cent over the previous year, and nonaerosol hair sprays had taken over nearly a quarter of the market, compared with 4 per cent in 1973.) It was believed that much work would have to be done to restore the marketers' confidence in aerosols.

There were, however, expressions of confidence that the aerosol would survive even if the fluorocarbons didn't. One economic study predicted recovery of the industry by 1980, by which time production should be "comfortably over the 1973 record [of nearly 3 billion aerosols]." The study said that even if all fluorocarbons were banned, production of aerosols could reach 3.2 billion by 1980 and more than 4 billion by 1985, because industry's ability to use propellants other than fluorocarbons "probably has been underestimated." (One is left to wonder, in light of this optimism, whether industry's protest about the inadequacy of other propellants had been little more than a political ploy to delay regulation as long as possible.)

Inevitably, the chlorine nitrate issue damaged the credibility of scientists. This was unfortunate, and perhaps a little unfair, but it was the unavoidable consequence of conducting public-policy science in a goldfish bowl. The scientists involved would undoubtedly have preferred it to be otherwise, but every move they made hit the headlines almost before they left the lab. The result was flip-flopping news stories that were almost bound to exasperate the layman.

Science News, for example, noted that "rumor is running rampant again . . . that the theory was wrong, industry was right, the ozone layer is saved, and fluorocarbons deserve an official reprieve. Or the inverse of all those statements, depending on who's talking." The *National Observer* ran a headline: "Remember that report that said spray-can gases don't threaten earth's ozone shield? Well, its data were wrong. Stay tuned."

Sherry Rowland, and, to a lesser extent, the other scientists who had supported a ban on fluorocarbons, were put somewhat on the defensive by the chlorine nitrate issue. They were inevitably accused of having cried wolf. A case in point was an article by Edward Edelson, science editor for the New York *Daily News,* who suggested that scientists had not tried very hard to communicate the uncertainties in their calculations. "On the contrary, their position seemed to harden as industry fought back. . . . The atmospheric scientists who warned about ozone did not speak of uncertainties, even though they knew that uncertainties existed. If they had admitted the possibility of imperfection, perhaps they would have been less damaged when an imperfection was found." Edelson also took politicians and the public to task for demanding simple answers to complex questions. "It is no wonder that scientists often yield to the temptation of giving straight yes-or-no answers when they should say maybe."

Some scientists felt a certain damned-if-you-do-and-damned-if-you-don't frustration with this. People were usually complaining that they were always saying "maybe," that they were always two-armed scientists ("On the one hand . . . but on the other hand . . ."). Some also felt that the charge was a little unfair—at least as a blanket charge against the entire scientific community—since the National Academy panel had set itself the precise goal of quantifying the uncertainties in the calculations.

In point of fact, the article seemed in some respects directed specifically at Sherry Rowland and others, like Ralph Cicerone, who were implicitly referred to as "alarm-raising scientists." To be fair, Rowland and others had always emphasized that there were uncertainties in the calculations, although at times this could be overshadowed by their insistence that the uncertainties were small enough, the risks large enough—and spray cans unimportant enough—that action should be taken immediately. One could argue that these scientists were in this case acting as no more than knowledgeable private citizens who had made personal value judgments and were exercising their right to engage in public debate on a matter of public policy; but it is probably naïve to suggest that their opinions, personal though they may have been, would not carry disproportionate weight in view of their special place in the scientific debate.

For his part, Rowland apparently decided that the best defense was an offense. He told *Food and Drug Packaging* magazine that industry should not pin its hopes to the chlorine nitrate reaction. "The bottom could fall out from under them very soon if they persist." Described as "ever-confident" by *Chemical and Engineering News,* he is quoted as saying: "I'm still recommending an immediate ban."

His position was based in large part on the climate problem—something that was initially and most vigorously raised by Paul Crutzen. The climate problem stemmed from the fact that chlorine nitrate caused a rearrangement of the relative amounts of ozone in the upper and the lower stratosphere. The formation of chlorine nitrate in the lower regions would protect ozone; higher up, where chlorine nitrate was more readily destroyed by solar radiation, it would liberate its chlorine oxide, and heavy ozone destruction would occur. The effect would be a lowering of the point of maximum concentration—or the "bulge"—in the ozone layer.

We saw in Chapter Two that in the process of absorbing the solar radiation that splits it apart, ozone controls the temperature of the stratosphere. Redistribution of ozone would therefore almost certainly affect the climate of the stratosphere. But scientists did not know what that would do to the climate at ground level.[3] The "coupling" beween the stratosphere and the troposphere—whereby what happens in one region can affect what happens in the other—is still not well understood by meteorologists. However, some mete-

---

[3] Some indication of what might happen at ground level was given in a computer study done by Ruth Reck, a physicist with General Motors Research Laboratories. She concluded that rearranging the "ozone profile" produced much larger surface-temperature effects than even near-total *removal* of the ozone layer.

orologists believed that the major climatic impact of fluorocarbons would result from factors other than ozone redistribution.

Nevertheless, the emergence of climate as perhaps a more important factor than skin cancer galvanized Paul Crutzen into action. Whereas before he had been rather lukewarm on the subject of immediate regulation, he now became almost adamant. (This rather puzzled others who had retreated somewhat because of the uncertainties that chlorine nitrate had introduced into the picture.) He is quoted in *Science News* after the Boulder modelers' meeting as saying: "I am sure that this industry will be phased out—if not for biological reasons, then climatological ones." Shortly after, he told another reporter: "There's no way this industry is going to survive for a very long time." Letting his computer model run out to the future had shown him "an atmosphere which is very different from ours at the moment. You'd think you are looking at a different planet. We cannot allow it to go on for long."

Mike McElroy was critical of other scientists for switching their emphasis from skin cancer to climate, seeming to regard this as little more than a political fallback position. "That wasn't a smart thing to do," he says. "If you want to build a case on the ozone [depletion] issue, then, when it disappears, you cannot suddenly switch and say you've got to ban anyway because of this other thing." He said other scientists "were covering themselves in a way that was reprehensible." They should have "hidden and ran" from the press, or discussed the situation "honestly and openly. The honest statement at that point was, 'We have a new problem, we cannot assess it with present technology, and we're developing new technology to assess it.'"

Rowland denies that he "suddenly" switched. He says he had always considered the climate problem to have a much greater "potential for disaster" than the skin-cancer problem. Nevertheless, he feared that others would judge the question of total ozone loss to be more important than the redistribution of the ozone with altitude. He feared, in particular, that if the National Academy study came out with this view, "it was going to be very hard ever to change that answer.

"I had a distinct feeling at that point that we could be real losers. . . . I felt that although we were basically correct, we were going to lose in the public-relations end of it, because it would be judged that the [total] ozone column effect had disappeared on the basis of chlorine nitrate. I had a period of twenty-four hours there where I was analyzing to myself what that would do, in the sense of having been publicly judged to be wrong. Eventually I rationalized my way out of it—that it would become clear eventually, but it might take ten years or so, to realize that the ozone depletion at higher altitudes was quite important."

The controversy over chlorine nitrate bubbled on through the spring and summer of 1976. Rowland, Molina, and Crutzen argued that chlorine nitrate had reduced the estimates of ozone depletion about a factor of 2; Cicerone, by a factor of 2 or 3. By the end of May, Rowland was down to a factor of 1.5, and he was increasingly dismissing chlorine nitrate as a major problem. He told the trade magazine *Food and Drug Packaging* that "we have revised our original theory to include the chlorine nitrate reaction . . . but we believe it will make no real difference. Measurements find no amounts that could substantially prevent ozone depletion."

This wasn't strictly true. Rowland was referring to an attempt by University of Denver atmospheric physicist David Murcray to measure chlorine nitrate in the stratosphere. Murcray was unable to detect its presence, but the sensitivity of his measuring technique was not good enough to permit unequivocal conclusions to be drawn from the data. The upper limit he set on the concentration of chlorine nitrate was just at the edge of what the modelers were predicting. However, his results were interpreted by Rowland (and by Philip Hanst of the Environmental Protection Agency) as indicating lower, and therefore significant, limits than Murcray himself was prepared to suggest.

Ironically, in at least one case, it appears that Rowland and Molina used arguments against the seriousness of the chlorine nitrate problem that were strangely reminiscent of the arguments that the industry had been using against their theory all along. Rowland and Molina told the Chicago *Tribune* that the chlorine nitrate factor "has never been proved outside the laboratory and has not yet been found in the stratosphere. . . ." Chlorine nitrate, they reportedly said, "is not significant; it just isn't an important factor." Most scientific colleagues considered it to be "of relatively little scientific interest" (a comment that might have amused the members of the National Academy panel who were still frantically trying to revise their report to account for chlorine nitrate).

A spokesman for the Aerosol Education Bureau retorted that Rowland's reaction was predictable "sour grapes."

By midyear, the combatants, like glaciers, had once again retreated to their respective poles. At a seminar on June 30, 1976, Mario Molina said: "There's no longer a basis to think we have a serious problem with the theory. . . ."

Du Pont's Peter Jesson responded: ". . . the theory is open to more question today than it was a year ago."

If the public confusion over chlorine nitrate continued, the Academy Panel, at least, was making considerable progress in resolving the issue behind the scenes. Things were turning out not to be as bad as the panel members had feared after the Boulder modelers' meeting. One happy development was that the modelers were at last beginning to resolve their differences. It turned out that these had been primarily due to the fact that some modelers had used a sun that shines with half its strength all the time, rather than a full-strength sun that turns off and on once a day, as the real sun in effect does. This technique makes life easier for the computer modelers, and it is an approximation that is usually acceptable, but not in the case of chlorine nitrate. (Sherry Rowland once complained, "It is unfortunate that the shortcomings of such models in simulating the stratosphere have not been as widely publicized as have the conclusions from them about total ozone depletion.")

As the work progressed, it became evident that the effect of chlorine nitrate was not going to be as large as originally suspected. At the time of the Boulder meeting, scientists did not have good measurements for the speed of the reaction that formed chlorine nitrate, and conservatively high values were chosen to put into the models. But the laboratory chemists had responded rapidly to the need for new data, and within a month three independent groups had measured the speed of the reaction. The measurements agreed with each other remarkably well.

With the new data, the various computer models indicated that chlorine nitrate would reduce the estimates of ozone depletion by about a factor of 1.5 to 1.8. The panel ultimately concluded that the effect of chlorine nitrate would be to reduce the fluorocarbon problem by a factor of 1.75.

By September 1976, the panel had modified its report to include the effect of chlorine nitrate; as suspected, the uncertainties in their conclusions had indeed been increased. This, combined with the fact that they had been caught completely off-guard by chlorine nitrate, made them extremely wary, and the tone of their final report was very much softer than it might otherwise have been.

In general, the panel concluded that the fundamental features of the Rowland/Molina theory were substantially correct. Most of the fluorocarbons that had been released were still in the atmosphere, and they were slowly rising to the stratosphere. There the chlorine they contained was being converted, through the absorption of ultraviolet light, to the reactive forms that catalytically destroyed ozone.

The panel also attempted to analyze the magnitude of the effects —the amount of ozone depletion that would occur. They used three scenarios: continued growth of fluorocarbons at 10 per cent a year, constant release at the 1973 levels; and a complete ban by 1978.

Continued growth at 10 per cent a year—the pre-1974 average— seemed totally out of the question. It would lead to a major disruption of the ozone shield in a matter of a few decades. (This was perhaps not too surprising, since we have come to recognize that exponential growth of anything leads rapidly to disaster.)

The second scenario—constant release at the 1973 levels—was one that some in the fluorocarbon industry hoped was feasible. The important question for the panel here was: What would be the globally averaged ozone reduction when the system reached steady state, and how long would it take to reach steady state? The panel concluded that ozone reduction would most probably reach 7.5 per cent at steady state,[4] but the uncertainties were such that it could be

[4] In June, Sherry Rowland had predicted that the panel would most likely settle on an ozone depletion of between 5 and 9 per cent.

as little as 2 per cent or as large as 20 per cent. This uncertainty was to a large extent influenced by the uncertainties surrounding the role of chlorine nitrate. If chlorine nitrate turns out not to be as important as assumed, the most probable value for the ozone depletion would rise from 7.5 per cent to as much as 14 per cent. (As we shall see later, more recent studies have produced numbers which are higher than the Academy's by about a factor of two; a study by NASA, for example, gives a central value of about 16 per cent.)

The panel also concluded that ozone depletion would continue to occur even if fluorocarbons were banned in 1978. Because most of the chemicals already released had not had a chance to get into the stratosphere, the concentrations in the stratosphere—and therefore the ozone depletion—would increase for a period of about ten years. The ozone shield would then start to recover, but this recovery would be extremely slow and the ozone layer would not return to the level of 1978 (when the ban was imposed) for about a century. Thus, the spray cans used during the 1960s and 1970s would leave a legacy of ozone depletion for perhaps four generations, into the mid or late 2000s.

The work of the Academy panel was finished. The members had weighed the scientific evidence and rendered their judgment. Their work would form the basis for the policy recommendations contained in the report of the Academy's Committee on Impacts of Stratospheric Change. All that remained was to release the information to the public.

In simpler days, with a less controversial subject, the rest might have been easy. In this case, as with just about everything else related to the fluorocarbon issue, it was not easy at all. The timing and manner of the release—the very wording the Academy used—were subject to as much controversy and competing pressures, and caused as much uproar, as any other aspect of the issue.

CHAPTER ELEVEN

No Oscars for the Academy

Scientists Back New Aerosol Curbs to Protect Ozone in Atmosphere.

—Headline in the New York *Times*,
September 14, 1976.

Aerosol Ban Opposed by Science Unit.

—Headline in the Washington *Post*,
September 14, 1976.

One could be forgiven a mild confusion in reading these front-page headlines in the *Times* and the *Post* the morning after the release of the National Academy study on fluorocarbons. Despite appearances, however, the most telling thing about the headlines was not that they were 180 degrees apart, but that they were both fundamentally correct.

The headlines and the attendant press accounts concerned themselves almost solely with the report of the Academy's Committee on Impacts of Stratospheric Change (CISC), rather than the lengthier and more technical report of the Panel on Atmospheric Chemistry. Both documents were released simultaneously at a press conference in Washington on September 13, 1976, but the CISC report inevitably attracted most of the attention since it addressed itself to the issue that represented the political bottom line: the question of regulation. (As we shall see later, the Academy would be strongly criticized for doing so.) It was the report's treatment of this delicate subject that led to the seemingly contradictory headlines. While stating that regulation was "almost certain to be necessary at some time . . . ," it also explicitly recommended against immediate regulation, saying that a delay of up to two years—but definitely no longer—was tolerable. In that time, some key uncertainties could be cleared up.

Nature magazine described the report, perhaps too generously, as "finely balanced"; others, like *Chemical and Engineering News,* described it less kindly but perhaps more accurately as "waffling." One reporter at the press conference was overheard to mutter that it appeared to be the work of four-armed scientists (a reference to complaints by politicians about two-armed scientists).

There is some irony in this turn of events. From the beginning, many people had been worried about the precise wording of the Academy's report and the impression it would leave on the public—so much so that this worry seemed at times to overshadow concern about the report's scientific content. This was undoubtedly a reflection of the fact that news stories—or, more probably, headlines—rather than the total report itself would determine what the public would learn about the Academy's study. Many scientists no doubt wish it could be otherwise, but this is the way it is. Therefore, before discussing the details of the report, and the reaction to it, it is useful to examine what went on behind the scenes prior to its release.

The industry showed an early and continuing interest in the wording of the Academy report and a preoccupation with the impact it would make on the regulatory agencies and the public—understandable in view of the tacit consensus that the Academy would be the Supreme Court on the issue. These concerns were reflected in a lengthy letter sent in January 1976—several months before the scheduled release of the report—to the Academy's president, Philip Handler, by Ted Cairns, director of Du Pont's Central Research and Development Department and a member of the Academy. Cairns said that the wording of the IMOS report put constraints on the wording the Academy could use in its report. IMOS, remember, had concluded that there was "legitimate cause for concern" and that if the Academy confirmed this assessment, federal regulatory agencies should begin rule-making procedures to implement regulations. Cairns said the only way he could conceive of the Academy not finding "legitimate cause for concern" would be to disprove the ozone-depletion theory—"something we judge to be impossible given the scientific facts at hand today." Thus, he said, if the Academy didn't deal explicitly with the IMOS statement, "a flood of restrictive legislation and regulation" might well follow the Academy's report.

Cairns went on to argue that basing legislation on an "unproven theory" would set an undesirable precedent and it might set the stage for future decisions based on even less well-established theories. "This is the road to the rule of witchcraft where, by definition, the accusation proves the charge."

Cairns further argued that if regulations were brought in that later proved to be premature—and if there were social and economic dislocations that proved to be unnecessary—this might have a strong negative impact on the public and "it might be a severe setback for the environmental movement generally. . . . If the fluorocarbon/ozone issue proves to be a 'cry wolf,' it will probably be the loudest and shrillest on record."

He concluded by saying that the report was quite likely to be used as the basis for legislation and regulation and that "its implications will be interpreted by many widely diverse groups. I would hope, therefore, that the Academy would address the public-policy questions directly by interpreting the significance of the report to provide guidance to the nonscientist policymakers."

In response, Handler wrote a short and politely noncommittal letter to Cairns, saying that he would share these ideas with those who were preparing the final document. Clearly mindful of the delicacy of the subject, Handler concluded by saying: "It is not appropriate that I respond here to the particular points that you make. Hence I trust that you will understand the lack of further comment."

Many scientists, including several referees of the panel report, were similarly concerned about questions of tone and wording, and we have already discussed Sherry Rowland's anxieties about how the chlorine nitrate issue would be handled. In view of the sensitivities all around, the Academy was particularly concerned about loose talk by panel or committee members and—with reporters hovering like vultures—about the possibility that the study's conclusions might be prematurely leaked to the media. When it happened, it naturally happened at the worst possible time—in mid-March, just at the point when the chlorine nitrate problem and Lazrus' HCl data were beginning to seriously shake things up. We have already seen that these new data had thrown the panel into a state of disarray. Their alarm was shared by the committee and by the Academy itself. Although the full impact of the new data had not been fully assessed, they felt that they had only barely avoided making a terrible mistake and there was an urgency to scramble back to firm ground as quickly as possible. This is why the apparent leak of the report provoked such a dramatic reaction—some would say overreaction—on the part of the Academy.

On March 18, Joel Shurkin made one of his semiannual pilgrimages to Pennsylvania State University. Shurkin, the science writer for the Philadelphia *Inquirer*, is an able and irreverent reporter who does not seek confrontations but does not walk away from them either. A former U.S. correspondent for the British news service Reuters, he once engineered a mutiny of Reuters reporters in Houston during the Apollo 13 moon mission in which three astronauts ran into difficulties in space that threatened their lives. The U.S. wire services, Associated Press and United Press International, were writing gloom-and-doom stories about the imminent death of the astronauts, and Reuters' head office in London wanted to know why its Houston team wasn't matching the AP and UPI stories. Shurkin and his colleagues didn't think the extreme doomsday angle was justified. Threatening resignation, they fired off a telegram to London:

It is not generally felt we should strive to be wrong in doubtful company. Feel Houston team must tell it like it is and cannot sacrifice principle of accuracy and honesty in representation for the sake of getting into a scare-story war of nerves between the

American agencies. . . . Associated's latest lead is already a running joke here with New York *Times,* Washington *Post, Daily News,* and other responsible papers.[1]

Shurkin was not sorry to leave behind the unremitting pressures of wire-service duty for the relative calm of daily newspaper work with only half a dozen deadlines to meet a day. He had been with the *Inquirer* for about three years in March 1976.

On the eighteenth, he went to see Hans Panofsky, a meteorologist at Penn State, who was at that time a member of the Academy's Committee on Impacts of Stratospheric Change. However, Shurkin was not after a story on the Academy's fluorocarbon study. He did not know that Panofsky was a member of the committee and he was interested instead in the scientist's work on clear-air turbulence. When Shurkin first called to set up the interview, it was Panofsky who volunteered to talk about the Academy study—something he later ruefully described as a "stupid thing to do." But he says that he really only wanted to tell Shurkin about the role and structure of the Academy committee—which was public information. When Shurkin arrived for the interview, he remembers Panofsky saying: "I guess you're here to talk about the ozone report." He wasn't, but the reporter in him told him to keep his mouth shut and listen.

In the course of their conversation, Shurkin asked if there was a draft of the committee report. Though Panofsky found this curious —he later conjectured that Shurkin *knew* there was a draft and this was a way of entrapping him into talking about it—this was not an illogical question for a reporter to ask when the official release date of the report was only a few weeks hence. Not wanting to lie, Panofsky said there was a draft, but he said it was preliminary and still subject to scientific review. According to Shurkin, Panofsky said that the Academy had determined that the fluorocarbons were indeed a threat to the environment and that there would be a recommendation to eliminate nonessential uses of the chemicals. (In his story, Shurkin said the *panel* would make this recommendation, but

[1] After a tense three days in space, the astronauts nursed their crippled spacecraft home safely.

he acknowledges that he was confusing the panel and the committee here. It was not the panel's mandate to make such a policy recommendation, and it never at any time contemplated doing so.)

Panofsky denies telling Shurkin that the report contained a recommendation on regulations. He says he tried to make clear that he was expressing a personal opinion when he said he thought the report, when it finally came out, would suggest regulations. This, he has since concluded, was "a naïve thing to say. I didn't tell him what the conclusions were, but I guess it was implied in the way I talked. . . . He could have figured it out. . . ."

Then Panofsky said, "I told him a little too much about something that *was* in fact in the report." He said that if regulations went into effect, they would most likely be selective.

However, Panofsky insists that he warned Shurkin that there was new chemistry on the horizon—he was referring to Rowland and Molina's chlorine nitrate work—that might "cause trouble before the report ever came out."

Shurkin says he asked Panofsky during their conversation if he preferred not to be named as the source of the information. Panofsky, he said, urged him to call other scientists associated with the study to corroborate his story. "I protested once or twice but he seemed to want to make sure I was not being misled," Shurkin recalled later. Panofsky denies urging Shurkin to call others.

Shurkin acknowledges that Panofsky did mention new findings, "but he left me with the distinct opinion that it was not something he took seriously." During the course of the interview, Panofsky was interrupted by a phone call, and Shurkin took the opportunity to write down some names from the list of scientists associated with the study. He was not purloining secret information here—the membership of the Academy panel and committee was public knowledge—but it would save him the trouble of trying to get the names through the Academy.

Shurkin left Panofsky's office at about three in the afternoon. He had a deadline approaching, and he went to Penn State's information office to place a call to his editor. Then he called the National Academy's information office. Howard Lewis, the Academy's information director, was not available for comment, and Shurkin left the Penn State number where he could be reached.[2] He asked media-relations director Barbara Jorgenson, who took the call, where the Academy had gotten the funds for the study, and she told him the name of the sponsors.

March 18 rapidly became what she later ruefully called a "fun day" at the Academy. Shortly after Shurkin's call, she got another call from a Washington journalist, who told her that a "major eastern newspaper" had obtained a copy of the report and would publish a story on it the next day. But the reporter would not say who it was. When she got off the phone, Jorgenson ran down to Lewis's office to tell him. Together, they went back over her phone messages. At that point it occurred to her that Shurkin's was a strange call. "He had not asked me the right question. He was the only person who didn't ask when it [the report] was coming out." So Jorgenson called Shurkin back and asked if he was writing a story on the report, and Shurkin confirmed that he was. Jorgenson went back to Lewis and said: "I think you'd better go upstairs" ("upstairs" being the office of the Academy's president, Philip Handler).

Meanwhile, back at Penn State, Shurkin had started calling the scientists on his list. One of them was Jim Friend at Drexel University in Philadelphia, who was a member of both the panel and the committee. Shurkin still had the two groups confused, and Friend tried to straighten him out on that (as it turns out, without success). But Friend would not comment on the panel report, refusing to confirm or deny anything Panofsky had said. As Friend remembers the conversation, he was "trying like hell" to avoid giving Shurkin any information. He kept repeating that he wasn't in a position to verify anything. Nevertheless, he says it is possible that he might have said something that Shurkin "construed as at least not denying what Hans had said."

Shurkin then filed his story on the basis of what he had gotten

[2] He later acknowledged this as an error on his part, for it allowed the Academy to identify Panofsky as the source of the leak. This was Shurkin's only regret about the incident. In retrospect, he feels he should have ensured Panofsky's anonymity, despite the scientist's willingness to be identified, because Shurkin believes that Panofsky really did not understand the kind of trouble it would cause. Shurkin did not identify Panofsky in his story, but he did use his name as a reference when he called other scientists, and this was another way Panofsky's name got out.

from Panofsky, but Shurkin told his desk to hold it until he could confirm it with a second source. This he was able to do in surprisingly short order; the next name on his list provided the confirmation. He called his desk and told them to go with the story.

Just before 6 P.M., with the first edition of the paper almost irretrievable, Shurkin got a call at his hotel from Howard Lewis. Lewis was, according to Shurkin, very upset by the apparent leak of the Academy study, particularly since the flap over chlorine nitrate and Lazrus's HCl data had, by that point, thrown the entire study into a state of disarray. Lewis said there was new evidence that had changed the whole picture, but he would not tell Shurkin what the new evidence consisted of.

Unpersuaded, Shurkin refused to kill the story. However, he called his desk and had them insert a paragraph quoting an "angry" Lewis as saying "There have been two major findings in the last few days that prompted the panel to withdraw its report." Shurkin added that "Lewis would not say what the findings were or how they might affect the report."

Shurkin's editor told him to double-check this with Panofsky. Shurkin tried to reach the scientist at his office but found he had already gone home. He tried again later, in the middle of his dinner, and found Panofsky at home. Panofsky, he said, mentioned the new chemistry, but "left me with the impression again that it was not likely" to cause a reversal in the report. It is clear that Shurkin simply did not feel himself warned off, while Panofsky expresses mystification that Shurkin could have gone ahead with the story despite the warning.

While all this was going on, the Academy grapevine had been working at full speed. Jim Friend received a call from panel chairman Herb Gutowsky demanding to know what was going on. Gutowsky did not approve of media interest in the panel's deliberations at the best of times—he once exasperatedly asked a reporter who questioned him if it was proper to approach a "jury that has a case under decision"—and this incident only served to increase his disapprobation. Friend recalls that Gutowsky spoke in a "very accusatory manner. He was somewhat like a parent who was trying to discipline his unruly child." For some reason, Gutowsky and others had apparently concluded that, since the Philadelphia *Inquirer* was

involved, Friend must be the culprit because he lived in Philadelphia. The illogic of this in an age of the ubiquitous telephone
seemed to escape them.

Friend protested that he had told Shurkin nothing about the report, and Gutowsky accepted his word. The Academy later put out
a memo saying that the source of the information had not been a
panel member. (Shurkin, however, says he did talk to at least one
panel member, whose identity he will not reveal.)

After receiving the call from Gutowsky, Friend then tried to
reach Shurkin at the *Inquirer*. Failing in that goal, he spoke to the
national editor, telling him that the story was premature, that it
would be a disservice to the Academy and the scientific community,
and that it would make the panel's job even harder than it was.
The *Inquirer*, however, was not persuaded that the story was
wrong, and the editor told Friend they would go with it.

Later in the evening, Shurkin resumed calling the names on his
list, and he says he talked to one more scientist who added considerably to what he already had. "In no case did I have to pull
teeth," he said. Friend was the only scientist of the four he talked to
who did not co-operate.

The final version of the story referred to three sources, and
Shurkin reported that they "made no mention of the report being
withdrawn or of any findings that would alter its conclusions."

The aftermath of the Great Philadelphia *Inquirer* Leak was as
lively as the event itself had been. According to Shurkin, the Academy launched an all-out effort to discover his informants. Panofsky's
involvement was already known; the contrite Penn State scientist,
flabbergasted when he saw the story, had called the Academy's president, Philip Handler, and the committee's chairman, John Tukey,
with apologies and an offer to resign "since I was doing more harm
than good." The Academy did not take him up on it.

Shurkin was glad to tell anyone who was interested that Friend
was not the informant, but he steadfastly refused to identify the
other two sources.

Herb Gutowsky was one of those who attempted to find out who
the sources were. Gutowsky seemed to take the incident as a reflection on his personal integrity, and when he called Shurkin shortly
after the story had appeared in the *Inquirer*, he was extremely irate.

Shurkin reports that Gutowsky opened the conversation by demanding to know who he could write to to complain about Shurkin's lousy reporting. The conversation went rapidly downhill from there, ending with attempts by Gutowsky to elicit the names of Shurkin's sources. Shurkin was by this time completely unnerved by the whole episode and by the "real meanness of it."

The *Inquirer* story started a chain reaction of events. The night Shurkin wrote his story, it went out on the Knight news service wire and was picked up by Bob Cooke, the science writer for the Boston *Globe,* who used it as the basis of his science column a few days later under the headline, "Reports Indicate Freon Is Culprit in Spray Cans." Not surprisingly, this elicited an immediate response from the industry. According to Cooke, industry representatives called his publisher to complain. In addition, A. H. Lawrence (who is with Du Pont's Freon Products Division but identified himself in his role as vice chairman of the industry's Council on Atmospheric Sciences) wrote a letter to the editor complaining that the Cooke story had omitted the Academy's disclaimers. Later, R. F. Stevens, a director of Du Pont, also wrote to the *Globe,* accusing it of relying on rumors and of being in an "indecent hurry to bury the fluorocarbon industry before it has been pronounced dead or is even running more than a slight fever. . . . Contrary to statements in the *Globe* article, we believe that it will be possible to resolve this issue through scientific study—but not in an atmosphere where emotional statements, unfounded opinions, and guesswork are given more evidence than facts."

Cooke stood by his story, the *Globe*'s *ombudsman* backed him up, and a note from the editor was published in response to Lawrence's letter saying: "Mr. Cooke and the *Globe* stand by his story, which did include the industry's view." It added that a paragraph giving the Academy's comments had been deleted from the story for lack of space, "but the deletion did not change the story's facts."

Next came a story in *Business Week* magazine. Discussing the leaked version of the Academy report, the article said: "The ax that has hovered over the $500 million fluorocarbon industry seems ready to deliver its deathblow." The story went on to say that some users of fluorocarbons "have already handed down their 'guilty' verdicts" and others "are preparing to abandon the product." Noting that Du Pont had switched its fluorocarbon technical staff to study-ing alternatives, *Business Week* suggested that "Du Pont is girding for what it sees as an inevitable ban."

Du Pont was mightily upset by the news stories and clearly went to a fair bit of trouble to stamp out these brush fires. This was reflected in a letter Ted Cairns wrote to the National Academy's president, Philip Handler, shortly after the leak. It was clear from the letter that Du Pont had actually called several of the reporters who had done stories to ask who their sources were. While no names were divulged, Cairns indicated that these journalists had confirmed speaking to members of the Academy committee. "We appreciate that what reporters say may sometimes include distor-tions, just as what they write may also be a little distorted," Cairns wrote, implying, if not actually saying, that the reporters may have been lying in defending their stories. Nevertheless, he went on, his concern about committee members talking to the press before com-pletion of the study "has been clearly deepened by this particular episode. I find such behavior irresponsible in the extreme and won-der if there is anything that can be done with existing and future Academy study committees with respect to reinforcing the urgent need for confidentiality and for restraint in the expression of opin-ion to the press or public during the study period."

Cairns said that the *Inquirer* story had had a ripple effect "with a rather strong negative impact." Du Pont was particularly upset by the *Business Week* article, which, Cairns said, misinterpreted Du Pont's research into fluorocarbon alternatives as "girding for a ban" because of a belief that the Academy had found fluorocarbons guilty. "This has led some of our customers to question our mo-tives," Cairns wrote.

According to one source at the Academy, the industry (particu-larly Du Pont) was concerned that if the Academy did not deny the stories, they would be confirmed by default. The Academy had, in fact, been trying very hard to keep the lid on things. The day of the *Inquirer* story, anxious phone calls went out from the Academy to the scientists associated with the study, telling them to avoid the press at all costs and, if they were trapped by a journalist, to refer all queries to the Academy's information office. For its part, the in-formation office issued statements about the new data that can only be described as cryptic and evasive. The statement on March 19, the day of the *Inquirer* story, for example, read:

> We very much regret the public discussion of a preliminary draft report of a study now in progress—the study by the Panel on Atmospheric Chemistry of the effects of fluorocarbons upon stratospheric ozone. The published discussion is premature and particularly unfortunate because, quite recently, the panel has received several pieces of new information that may bear upon the conclusions of the study. Assessment of this new information is under way. Completion of this evaluation and its incorporation into the report will delay release of the report by at least a month.

According to Academy insiders, the Academy had come under considerable pressure to "do something." Cairns reportedly phoned the Academy's information office; then, deciding it was "not on the right floor," called Handler instead.

On April 5, the Academy issued a lengthier, if no more enlightening, statement saying that "newly received information, now being analyzed and evaluated, has delayed by one or more months the scheduled issuance of [the] report." Throughout the uproar, the Academy refused to identify the chlorine nitrate problem and the HCl data as the "new information." The Academy's reticence on the subject puzzled members of the scientific community; many felt it was a counterproductive tactic that simply fueled speculation and created rumor and confusion. Intentionally or not, the Academy left the impression that it was sitting on some new bombshell that was going to turn the whole fluorocarbon controversy on its ear. In its clumsy efforts to neutralize the Philadelphia *Inquirer* story, the Academy had plugged the leak with a hand grenade.

Howard Lewis, the Academy's news director, explained the reason for this: The new data were unpublished, and the Academy felt it had no right to pre-empt other scientists by publicizing or reporting on their work in progress. "I think in retrospect that it was probably a wrong decision to make, but we felt here that the right of publication is owed to the scientist." The ironical thing about this is that during this period, the scientists doing the work were not only busily discussing the new data among themselves, they were also openly

fielding queries from dozens of reporters. Before long, numerous stories appeared that dealt with the new data in considerable detail
and inevitably conveyed the impression of an evasive and slightly
hysterical National Academy. Ruefully, Lewis acknowledges that he
wasn't at the time aware that "there was this great network of people calling everybody else. If I had been, I would've argued
[against the Academy keeping quiet], but I didn't realize that everyone was talking to everyone else. In view of the fact that the
next issue of *Science News* that I saw told me not only all the things
I was trying to keep secret but twice as much as what I knew, I felt
kind of foolish."

According to one Academy source, there may also have been, at
least initially, a touch of paranoia in the Academy's response.
"There were a lot of people who wanted to get their hands on the
report and there were rumors floating about at various times that
people had them." That anyone on the panel or the committee
would actually tell a reporter what was in the report—particularly
when the conclusions were in such a state of uncertainty—seemed
inexplicable. The Academy source said it occurred to them that the
report may have been leaked by "someone who was pro-industry
and wanted to create confusion because new information would further obfuscate the question and industry could say these atmospheric scientists don't know what they're talking about—one day it's
a disaster, the next day it doesn't exist. There would be no credibility for the scientific community. So I think that was going on—
wondering whether there was some motive, believing there had to
be some nefarious motive." This particular conspiracy theory seems
implausible—one could not be sure that public opinion would not
solidify around the idea that the Academy had in fact found
fluorocarbons guilty—but it does demonstrate the siege mentality
that temporarily prevailed at the Academy.

There were also other pressures. For months, news stories had
been talking about the "long-awaited Academy report," saying that
everything was hinging on it. This impressed the Academy with the
need to think out carefully exactly what it wanted to say before it
went public. Suddenly it was forced to go public "and we weren't in
a position to deal with that," said Bruce Gregory, executive secretary of the committee and the panel. The uncertainties associated

with the new data meant that even the Academy didn't know what the conclusions would ultimately be. "We didn't want to create in the public mind the idea that there was more confusion than in fact there was."

An added difficulty at this stage, as we saw in Chapter Nine, was the fact that the government agencies funding the study were reluctant to shell out more money, and it was uncertain whether the project would be completed at all. All of these factors caused the Academy to adopt a "batten down the hatches" profile, at least in the short run.

Behind the scenes, the committee continued to grapple with the question of regulating fluorocarbons. Prior to the chlorine nitrate flap, the committee had indeed been hawkish on the subject, prepared to recommend restrictions. But the chlorine nitrate episode caused them to retreat and, having been burned once, they were less inclined to leave themselves wide open again. A draft of their report, reflecting this new caution, went to the scientific referees, and it soon ran into trouble. Some of the referees felt that the committee had backpedaled too much, that it was now too soft on the question of regulation. Moreover, there were complaints about the style of the report, which had been written largely by the committee's chairman, John Tukey, an expert in statistics from Bell Laboratories, Princeton University. It was dominated by Tukey's rather florid prose, which never used a single word where half a dozen heavily qualified phrases would suffice. The wrangling that ensued to get the committee report into shape introduced a further element of delay into the Academy's timetable for releasing the study. This was not welcomed by anyone, least of all the Academy, which was still being leaned on by various agencies to get the report out. Adding to the pressure was the fact that an international scientific meeting on the stratosphere was going to be held in Logan, Utah, on September 15, 16, and 17, and many scientists wanted the report available for discussion there.

These delays and pressures had a direct bearing on the Academy's strategy for releasing the study. Originally, they had planned an elaborate two-step procedure that called for the panel and committee reports to be released separately, a few weeks apart. This would allow time for the scientific community to comment on the panel report before the committee issued its recommendations. This plan was conceived during the pre-chlorine nitrate days, when the panel had such a strong report that it was, in the words of one Academy official, bound to "draw fire." The Academy thought it would be a good idea to "let the smoke and the flames clear" before committing itself publicly to policy recommendations. In addition, Howard Lewis also saw the staged release as a good way of coping with the delays in the committee report while appeasing the demand for the panel report to be ready for the Utah meeting.

Science writers, however, were less than thrilled with the plan. It was logistically messy, for one thing. For another, they didn't want to wait several weeks for the Academy to draw the bottom line. If the Academy intended to make policy recommendations based on the panel's findings, it should do so right away. In fact, the Academy began to worry that *no one* would wait around for the committee report. "We just couldn't have said, 'Stay tuned for a month,'" said Bruce Gregory. Remember that the IMOS report had stated that, if the Academy confirmed the IMOS assessment of the seriousness of the fluorocarbon threat, regulatory action should be started. The panel report would unhesitatingly be construed by the regulatory agencies as confirmation—as, indeed, it was. Releasing the panel report first "would have set in train a whole set of actions that would have made the Tukey report more or less anticlimactic," said Howard Lewis. He remembers discussing the matter with Tukey and Handler, saying that the committee could end up "talking to an empty house. Everybody will have gone. I don't think they'll be waiting for the other shoe to drop."

The committee did not find the prospect of talking to an empty house appealing, so with the deadline of the mid-September meeting in Logan looming, the pressure to whip the committee report into some sort of acceptable shape intensified.

The Academy made the deadline—barely. The two reports, as we have noted, were finally released together, on September 13, 1976, more than five months behind the original schedule. The panel report concluded: "It is inevitable that [fluorocarbons] released to the atmosphere do destroy stratospheric ozone." On the basis of this finding, the committee report stated: "It would be imprudent to accept increasing [fluorocarbon] use, either in the United States or worldwide." Selective regulation of fluorocarbons was "almost certain to be necessary at some time and to some degree of completeness." The committee felt, however, that measurement programs already under way promised to reduce the uncertainties in the calculations and therefore recommended a "strictly limited delay" of not more than two years to allow this research to be completed. The committee bought the argument that the penalty involved in such a delay would be small—that it would amount to an additional ozone depletion of no more than a fraction of a per cent. The bottom line in all of this came in the committee's recommendation No. 6: ". . . we wish to recommend against decision to regulate at this time."

In addition, the report recommended that future regulation of fluorocarbons be applied selectively to different uses at different times; that all products containing fluorocarbons not in sealed containers be labeled as an "aid to consumer self-restraint" in the use of the chemicals; and that the authority of existing legislation and regulatory agencies to control fluorocarbons be immediately strengthened.

The committee report also devoted considerable attention to the subject of the *direct* impact that fluorocarbons would have on the climate. We discussed in Chapter Nine the indirect climatic impact that results from reduction or redistribution of ozone.

The direct climatic impact of fluorocarbons has nothing at all to do with ozone; it has to do with the properties of the fluorocarbons themselves. Fluorocarbons are strong absorbers of the infrared radiation given off by the earth; by trapping this radiation and preventing it from escaping into space, they can therefore contribute to a warming of the lower atmosphere. This is the "greenhouse effect" and is identical to the trapping done by carbon dioxide in the earth's atmosphere.[3]

The possibility that the greenhouse effect might be an important factor in the fluorocarbon debate was first raised in the fall of 1975

[3] The concentrations of carbon dioxide in the atmosphere have increased about 10 per cent since the late 1800s, largely as a result of industrialization and the burning of fossil fuels. Some computer calculations suggest that these concentrations may double within the next fifty to one hundred years, with a consequent increase in the surface temperature of perhaps two to three degrees Celsius.

by atmospheric physicist Veerhabadrhan Ramanathan (then at NASA-Langley Research Center, and now at the National Center for Atmospheric Research). Using Ramanathan's work as a basis, the Academy committee calculated that if fluorocarbon releases were held to the 1973 rates, this would cause a warming of less than half that expected from carbon dioxide by the year 2000. As the report noted, however, the effects of fluorocarbons would be *"additive* to and in the *same* direction as the effects of CO_2" (emphasis theirs). The committee further calculated what would happen if fluorocarbons continued to grow at a rate of 10 per cent a year, which they had been doing up to 1973. In this case, their temperature effects would equal those of carbon dioxide before 2000 and would thereafter begin to dominate. These estimated climatic consequences, the committee concluded, were sufficiently large that continued release of fluorocarbons at the 1973 rates "must be recognized now as potentially serious." Continued growth of fluorocarbon releases—even if only a few per cent a year—could in a century or two lead to "climatic change of drastic proportions."

The committee report's style, as much as its content, had a significant public impact. Its "on the one hand, but on the other hand" character accounted for a certain ambivalence in the message that ultimately reached the public. (*Chemical and Engineering News* reported that the Academy had ". . . apparently sealed the fate of these otherwise innocuous chemicals—sort of.") In general, however, the media interpreted the report as a blow against the aerosol industry and confirmation of the Rowland/Molina theory. "Aerosol Peril to Ozone Is Confirmed" was a fairly typical headline. The prize for the most lurid headline goes to Sherry Rowland's local paper, the Orange County *Daily Pilot*, which spread "Aerosols Spewing Earth Death" across the top of its front page in Second Coming type. Interestingly, the *Morning News* in Du Pont's hometown of Wilmington, Delaware, carried the headline: "Ozone Study Gives Du Pont Reprieve."[4] The Detroit *Free*

---

[4] Reflecting the concerns of its own community, this news story began: "The Du Pont Company and its six hundred Wilmington-area employees involved in the fluorocarbon business can breathe a little easier today. But their worries over jobs and corporate income tied to the industry are not over."

Press ran a survey in which readers were asked if they were using fewer aerosols as a result of the Academy's finding that the Rowland/Molina theory was correct. About 89 per cent said "Yes" ("The environment is more important than the conveniences of an aerosol can."), and nearly 11 per cent said "No" ("They're a bunch of Communist crackpots and don't know what they're talking about.").

The industry was quick to make the most of the Academy's recommended delay in regulations. Ignoring the fact that the study had confirmed the validity of the Rowland/Molina hypothesis, COAS took out full-page ads that quoted, also in Second Coming type, the Academy's recommendation "against decision to regulate at this time." The recommendation is reprinted in its entirety, COAS said, "to prevent misunderstanding which may have arisen in the minds of the public" (presumably from news stories that said: "Scientists Back New Aerosol Curbs. . . ."). The ad goes on to say: ". . . we feel the National Academy of Sciences has been emphatic; there is not sufficient scientific evidence against fluorocarbons to regulate now."

Many industry representatives and some news stories suggested that, in fact, the committee report had essentially backed up the industry's campaign for a delay in regulations. The committee's chairman John Tukey, tried to keep arm's length from such an interpretation. He emphasized that the committee had recommended a delay of "no more than two years" and said that if the uncertainties were cleared up sooner, "then we favor regulation sooner." They were telling the political leaders to seek clarification, but if they hadn't found it by the end of two years, then they'd have to make up their minds anyway. "That's not the same thing as what industry was saying," Tukey maintained. The industry was, however, undeterred by the committee's reluctance to climb into bed with them. The day of the report's release, Du Pont issued a press release that ignored the distinction Tukey was trying to make and implied that the two-year delay was a hard-and-fast figure and not an absolute upper limit. It said that the recommendation against making a decision to restrict fluorocarbons *"while research continues for two years* was a difficult but correct decision. . . ."[5] (Emphasis added.)

[5] Just a few days later, however, Du Pont issued another Press release on the same subject, which contained the phrase "up to two additional years."

All of this was, however, mild compared to the reaction to the report of the Western Aerosol Information Bureau (WAIB), indefatigable in its determination to put an optimistic face on things. Walter Leftwich, head of the public-relations firm hired by the WAIB to oversee the industry's game plan in eleven western states, gave a speech about the NAS report. The NAS report, Leftwich said, had changed their game plan. "What we had done in the past was primarily defensive work. We were fighting brush fires, issuing denials, and correcting mistakes. Now, for the first time, with the issuing of the NAS report, we have been given the opportunity to move out aggressively, in a positive, hard-hitting program. We have the official report of the NAS to back us up. We have scientific quotes from a report to Congress to *confirm our position*." (Emphasis added.)

Leftwich said their first action was to get out a press release "that covered exactly what the NAS report said, and what it did not say!" And what *exactly* did the report say? "Stripping away the official gobbledygook, what the report says is, 'We don't know what is going on, we don't know if what we are measuring has anything to do with what the problem may be, but we're sure that nothing is going to happen one way or the other for the next couple of years, so don't make a move that will make us look more stupid later.' "[6]

If the committee report was less than a model of clarity and incisive argumentation, it nevertheless deserved better than this!

These industry tactics raised hackles at the National Academy. In fact, according to one source, the Academy's president, Philip Handler, was moved to protest to Du Pont and to write to members of the Academy that the industry had done itself a disservice.

Sherry Rowland had been on tenterhooks for three weeks prior to the release of the Academy report. With a slight self-deprecating chuckle, he remembers how he set up a tape recorder to catch the news broadcasts on the day of the report's release, so he could hear exactly what was said. Rowland felt strongly that his subsequent

[6] Sometime later, Schiff was asked to give a summary of the NAS report at an American Chemical Society meeting. After doing so, he told his audience he could have saved them time by reading this industry summary. He then read the summary, which was greeted by loud guffaws and considerable incredulity, even by industry scientists.

scientific reputation depended very much on what the media interpreted the report to say. "If they said 'Yes,' then we were clean. If they said 'No,' then we were hopelessly done in." The first broadcast he heard sounded like "Yes." The next seemed more equivocal. "By the third or fourth one, it sounded as though it was 'Yes, but . . .'" Rowland could live with that, even though the Academy's report had gone against his public-policy recommendation for an immediate ban. He felt he had given good advice, but his scientific reputation did not depend on whether he gave good political advice. "I was concerned about the public policy, but I was more concerned about my reputation, and my reputation doesn't depend on public policy—it depends on whether I'm right scientifically." The report had confirmed that.

Rowland feels, nevertheless, that taking a political stand has forever marked him as controversial and that "affects my reputation in the scientific community on a permanent basis." He does not regret taking that stand, but he says he did it recognizing that he had placed himself "in a group that is forever suspect."

(It may be true that Rowland will be "forever suspect" on public-policy issues. Having taken a strong, even partisan, position on the fluorocarbon issue, and having been identified with the environmentalist camp, his views on subsequent social/political issues might well be regarded with some skepticism by scientists. But Rowland is wrong if he believes that he is compromised as a scientist, because the scientific community, for the most part, makes a distinction between scientific and political activities. There were many who challenged Rowland's call for an immediate ban on fluorocarbons, but none who believed he was falsifying or distorting his data; indeed, he was commended for bringing the role of chlorine nitrate to light. His laboratory measurements were of a high caliber and were used by people on both sides of the debate. His work has not been refused publication in scientific journals because of the political stand he had taken, nor is it likely to be in the future. There is no question that Rowland could continue to function as a scientist; the real question is whether he wants to, or whether he finds the political/scientific battle a new and more challenging milieu.)

For public consumption, Rowland declared himself "gratified" by the report's confirmation of the hypothesis. He told the *Wall Street Journal* that restrictions on fluorocarbons should be—and probably would be—undertaken immediately. Initially, Rowland responded mildly to the recommended delay in regulation, telling his local paper, the Orange County *Daily Pilot,* that it didn't really bother him. Later, however, he began to speak out against it more emphatically.

There were some scientists associated with the Academy's study who feared that the committee report waffled so much that it would satisfy no one; what they did not anticipate is that it might satisfy everyone. (If this was the result of a deliberate decision, rather than inadvertence, perhaps the committee took a lesson from Henry Kissinger, who reputedly negotiated with the Arabs and the Israelis by telling each side exactly what it wanted to hear.) It certainly demonstrated the "finely balanced" (as *Nature* magazine described it) qualities of the report that Sherry Rowland was "gratified" while Du Pont commended the Academy for a "difficult but correct" decision. (They were, of course, talking about different aspects of the report.) *Newsweek,* however, put it more aptly: The report had left the dispute "almost as confused as ever—and both environmentalists and industrialists could claim minor victories in the decision."

It was not long, however, before industry's optimism was deflated. In light of the fact that the validity of the Rowland/Molina theory had been confirmed, the pressure for regulatory action began inexorably to mount. The Natural Resources Defense Council renewed its campaign to get the federal regulatory agencies to act. But the most dramatic developments occurred just three days after the Academy report was released. At the scientific meeting held in Logan, Utah, Russell Peterson, then chairman of the President's Council on Environmental Quality,[7] delivered a strongly worded speech calling for the regulatory process to begin immediately. "From the pure scientific perspective, there remain valid doubts about the effect of fluorocarbons on the ozone shield. From

[7] Peterson, who had once been a chemist at Du Pont, left the CEQ shortly after this to become head of New Directions, a citizens' lobbying group devoted to global environmental matters.

the public-policy standpoint, however, there remains no valid reason to postpone the start of regulatory procedures." He said it was necessary to "choose the less damaging of two alternatives. In comparison with the potential health-effects of uncontrolled fluorocarbon use, the potential economic losses associated with wise regulation are small. We must begin now to pay the modest costs of safeguarding the priceless health of our people and the place we live."

Recognizing that the regulatory process was a long one, Peterson also urged industry to voluntarily start phasing out fluorocarbons to minimize the ultimate dislocation. "It's clear that the benefits to the consumer in using fluorocarbons for underarm spray deodorants and hair sprays do not outweigh the threat to world environment from the continued use of such propellants. Hence, I recommend that consumers stop using them and manufacturers voluntarily stop selling them."[8]

A similar theme was struck by another speaker, David Pittle, of the Consumer Product Safety Commission. Pittle, however, sharply rapped the Academy committee on the knuckles for venturing into policymaking. He said the recommendation for a delay in regulations would make life difficult for the regulatory agencies; the recommendation carried added credibility simply because the Academy had made it, and regulators would have this thrown up to them if they tried even to start regulatory procedures. This brought up a second point: Pittle questioned whether the committee really understood the complexity of the regulatory process. It can take two or three years from the commencement of regulatory action to its final implementation—the result of public hearings and possible court challenges. (Thus, even if the process began right there, in September 1976, the regulations could not take effect much before

[8] During the question period following Peterson's talk, Hugh Ellsaesser, a scientist from Lawrence Livermore Laboratory, who had been pro-SST in the earlier ozone fight, got up to point out that the 7 per cent ozone depletion predicted by the Academy was the equivalent, in terms of causing skin cancer, to moving eighty-four miles South. After a moment's hesitation, Peterson bluntly said: "That's a silly argument." There was a momentary, stunned silence in the audience, then what sounded like a collective, stifled snicker. Finally, some in the audience, overcoming good manners, started to applaud Peterson.

1978 anyway.) Pittle expressed his conviction that the process should begin immediately; he particularly questioned the wisdom of delay, since it was unlikely that the additional study the committee advocated would result in a recommendation *not* to ban aerosol propellants. He did not believe regulations should be held up "merely to refine the precision of our assessment of the problem."

Later, Pittle got into a heated exchange with Fred Kaufman (a member of both the committee and the panel) that demonstrated the philosophical differences between the scientist and the regulator. Kaufman defended the committee's inclination to wait for more data: "I do not want to be associated with having made a stupid decision on the basis of insufficient evidence." Pittle countered that *he* didn't want to wait if the only result of additional data would be to underscore the threat already identified. "And I think that's all it's going to be." He added that he was the one who would have to answer the people who would ask why it took six years or more to enforce regulations after the problem was identified.

Pittle was not arguing that scientists shouldn't be allowed to express their views on public policy. Indeed, he urged them to make recommendations to the government. "Your technical recommendations are entitled to great weight. But your recommendations as to *policy* should not carry more weight than those of any other informed and responsible citizen." (Emphasis his.)

Predictably, the remarks of Peterson and Pittle provoked industry's wrath. They began waving the committee report as their New Testament—which made some of the scientists associated with the Academy study visibly uncomfortable. "No less than the full weight and prestige of the National Academy of Sciences lies behind its committee's recommendation that no decision to regulate be made at this time," said Du Pont. Peterson's statement "flies in the face of more than a year of intensive study" by the Academy, according to a Du Pont press release handed out at the Utah meeting, half an hour *before* Peterson gave his talk.

In the end, nothing about the report caused more confusion and dissension than the recommendation for a delay in regulations. It is problematical whether the Academy really needed to leave itself open to such criticism. The fact was that the regulatory agencies really didn't care very much what the scientists thought about regulation. Regulation was *their* job. All they wanted was a scientific assessment of the validity of the theory and the magnitude of the possible effects and they were prepared to carry the ball from there. The decision to regulate, Peterson said, was a social value judgment, not a scientific judgment—and he was right. The committee could have saved itself a lot of unnecessary grief by avoiding the recommendation for delay—a feeling shared by some of those who served on the Academy study.

John Tukey later explained why the committee made the recommendation. He said they felt it was the best way to convey to non-scientists, particularly those in Congress and the regulatory agencies, just what the uncertainties in the calculations meant. The committee didn't want simply to say that the uncertainties caused the predicted ozone depletion to range from 2 to 20 per cent, with a probable value of 7.5 per cent. "If you leave it just in numerical form, do you convey the information adequately," Tukey asked. "I didn't feel the committee's conclusions would be properly interpreted if we tried to say it in scientific language." And so they decided to say it in language that could be understood—the language of regulation. Another factor that came into play here was that the committee knew the regulatory agencies were champing at the bit. "I don't think anything less than the National Academy of Sciences saying this was scientific nonsense—and that we know ozone is not being reduced—would have stopped the regulations from coming," Tukey said later.

Bruce Gregory, the National Academy's staff officer on the fluorocarbon study, pointed out in addition that IMOS had really forced the committee's hand by saying that the regulatory process should begin if the Academy confirmed the validity of the Rowland/Molina hypothesis. The Academy report had done that; if it said nothing explicit about regulations, it would then be in the position of endorsing by default the IMOS recommendation that the regulatory process be started. If IMOS hadn't put the Academy in this position, it is possible that the committee might not have made an explicit recommendation concerning regulations.

The matter became moot less than a month later. IMOS unanimously recommended that the regulatory agencies commence action, and the Food and Drug Administration announced that it proposed to phase out all nonessential uses of fluorocarbons. The term "nonessential" was, at that time, a euphemism for deodorants and hair sprays.[9] Despite its efforts to portray such spray products as necessary and essential to the consumer, the industry simply could not overcome the overriding opinion that such products were really not worth taking a chance for. While one might think twice about abandoning refrigeration in the face of uncertain risks, it is not nearly so traumatic a decision to play it safe when it comes to spray cans. The New York *Times* put it this way: "To hold out for ultimate proof for the sake of these items would be to 'balance' a few more years of cosmetic convenience against the risk of losing a possibly dangerous percentage of the ozone essential to life on this planet." This theme was also reflected in the comments of Alexander Schmidt, then head of the FDA, who said in announcing the planned phase-out: "The known fact is that fluorocarbon propellants, primarily used to dispense cosmetics, are breaking down the ozone layer. Without remedy, the result could be profound adverse impact on our weather and on the incidence of skin cancer in people. It's a simple case of negligible benefit measured against possible catastrophic risk, both for individual citizens and for society. Our course of action seems clear beyond doubt." Schmidt said he could not justify further delays to allow more research to be done. Seeming to agree with Pittle that this further research would be nothing more than a fine tuning of what was already known, he said that a narrowing of the range of ozone depletion would not change the regulatory situation. "Given the effects on human health, even a 2 per cent depletion for "unessential" uses of fluorocarbons is undesirable."

Schmidt also announced that, as an interim measure, the spray cans would have to carry a label warning consumers that they contain a chemical "that may harm the public health and environment by reducing ozone in the upper atmosphere."

The FDA later embodied these measures in a formal notice of intended rule-making published in the *Federal Register*. In it, Schmidt announced that he had concluded that the information already at hand was sufficient to argue that fluorocarbon propellant

[9] As we shall see later, defining "essential" uses of fluorocarbons and deciding what to do about them posed greater problems.

use "poses an unreasonable risk of harm to the public health and the environment."

With respect to the warning label, he noted that the wording would minimize "any possibility that the consumer will believe that the warning refers to risks of harm from direct inhalation of the products."

At about the same time, the Consumer Product Safety Commission finally acceded to the Natural Resources Defense Council's petition to ban fluorocarbons in spray cans, and the Environmental Protection Agency also announced its intention to propose regulations.[10] In the interests of consistency, the three regulatory agencies formed an interagency work group, chaired by the EPA, to co-ordinate their activities. Russell Train, then the EPA chairman, estimated that restrictions on nonessential uses of fluorocarbons would limit demand for the chemicals to 47 per cent of the total U.S. consumption in 1975.

In some respects, these prompt actions by the regulatory agencies may have damaged the Academy's credibility, particularly with the industry making such a big splash of the Academy's recommended delay. Ironically enough, however, the agencies' proposed actions were not really inconsistent with the Academy recommendation when you consider, as Russell Train pointed out to a congressional subcommittee: "Only by starting now, can we expect to issue final regulations within the next two years." Nor was John Tukey averse to this. "We said don't decide to regulate; we didn't say don't do the year's work required to get to that point."

On May 11, 1977, in an unprecedented action, the FDA, the EPA, and the CPSC jointly announced the timetable for phasing out fluorocarbon propellants in spray cans.[11] The ban would take place in three steps: First, the manufacture of fluorocarbons for use as propellants would be prohibited by October 15, 1978; then by

[10] In October 1976, the EPA sent a notice to producers of pesticides, strongly urging them to use propellants other than fluorocarbons. This was to be a voluntary effort; however, failure to volunteer could result in cancellation of product registrations.

[11] By this time, annual spray-can sales were down from the 1973 peak of nearly 3 billion to less than 2.3 billion—about the level of 1968. Personal products were down 34 per cent from the peak. It was estimated that fluorocarbons were used in less than half of all spray cans and possibly as few as one third by 1977.

December 15, 1978, all companies would have to stop using existing supplies in aerosol products; finally, by April 15, 1979, all interstate shipments of products containing fluorocarbon propellants would be banned. If the ban goes into effect on time, the Spray-can War will have lasted five years.

The labeling requirements proposed by the FDA went into effect in the fall of 1977. The CPSC had proposed similar labeling requirements for products under its jurisdiction, and the EPA had already ordered labels on pesticide sprays containing fluorocarbons. (The label—warning consumers that the propellant "may harm the public health and environment by reducing ozone in the upper atmosphere"—was intended to encourage voluntary cutbacks by consumers. In the summer of 1977—in what was probably the first major TV ad campaign by a marketer to deal explicitly with the ozone controversy—the label was prominently featured in an ad for Ban Basic pump spray deodorant. In pushing the pump-spray alternative, the ad said "If you use an aerosol, you may be concerned about the order to ban fluorocarbon propellants in two years." It was, of course, inevitable that someone would eventually turn the ozone controversy into an advertising plus.)

Donald Kennedy, new head of the Food and Drug Administration, said of the proposed regulations: "We judge these actions to be a prudent and proper response to the problems of human health and safety imposed upon our world by a chemical introduced and used in innocence but now apparently both insidious and destructive."

EPA Administrator Douglas Costle said that the eighteen-month phase-out had been chosen because it would "permit an orderly transition to available alternatives without causing undue economic hardship to industry or the consumer." The EPA estimated that the switch to other propellants and packages would cost the industry $169 million to $267 million for each of the four years after the ban. However, since the alternative packages might be cheaper, consumers could save $58 million to $240 million a year if industry passed on its savings to the consumer. An industry spokesman was quoted in the New York *Times* as saying that they could meet the deadline, but it would produce "an economic hardship to industry and ultimately to the consumer."

This did not, however, seem to be universally the case. The next day, the *Times* ran a story headlined: "Outwitting Aerosol Ban: New Systems Ready." The story began: "The proposed government ban by 1979 on spray cans using fluorocarbon propellants seemed to create no great concern yesterday among major suppliers or users. On the other hand, the Precision Valve Corporation of Yonkers and the Selvac Division of Plant Industries, Inc., were rejoicing. Both say they have developed new propellant systems that would meet all government requirements." You will recall that Precision Valve's president, Robert Abplanalp was one of those most contemptuous of the Rowland/Molina theory. Although he insisted that the theory was nonsense, he had clearly been hedging his bets. Abplanalp's new spray system, called Aquasol, is designed to work with a nonflammable water-butane mixture. After the announcement by the regulatory agencies, Abplanalp said he still thought the ozone controversy was a "hoax" but that his new system "would have wiped out fluorocarbons anyway." *Time* magazine observed: "If Aquasol passes market tests and proves as popular as aerosol, the nation's medicine cabinets could remain stocked with spray cans a long time into the future."

Despite the ferocity of their campaign to save fluorocarbons, Du Pont officials, in the end, seemed to be trying to dismiss the ban as inconsequential to them. In the *Times* story, a Du Pont spokesman is quoted as saying that Du Pont sold about $250 million in fluorocarbons in 1976, about $50 million of that for aerosol products. Thus the fluorocarbons accounted for about 3 per cent of Du Pont's total sales of $8.36 billion for that year, with aerosols accounting for only .5 per cent. "It wasn't very significant," the spokesman said.

Since the new regulations were to apply to nonessential uses of fluorocarbons, it was necessary for the regulatory agencies to define what they meant by "essential." To the FDA, it meant three things: (1) that there were no technically feasible alternatives to the use of fluorocarbons in the product; (2) that the product provided a substantial health, environmental, or public benefit unobtainable without the use of fluorocarbons; and (3) that the use does not involve significant releases of fluorocarbons to the atmosphere, or, if it does, that the release is warranted by the benefit conveyed.

There were many products in addition to hair sprays and deodorants that the FDA did not consider essential; among them were perfumes, pan-coating cooking sprays, and, ironically, sunburn sprays. This, of course, involved value judgments—many of which were challenged by industry spokesmen, who came up with a long list of the beneficial qualities that fluorocarbons imparted to various products. (Examples: They did not alter perfume fragrances; they evenly distributed a minimal amount of pan-coating spray; they allowed medicines to be applied to sensitive areas without touching.)

The FDA was unimpressed with most of these arguments, particularly the ones for cosmetic uses of fluorocarbons. Many of these products were available in alternative forms and thus, while the sprays might offer some advantages in terms of convenience, this was not deemed sufficient to class them as essential. Nor did medicines and pharmaceuticals get an automatic exemption. Among those that did not were topical analgesics, sunburn and minor-burn remedies, topical antimicrobial and antifungal agents, surgical scrubs, a surgical-spray dressing, and insect-bite products. Prescription drugs, the FDA said, would have to be considered on a case-by-case basis; some drugs are in cans for convenience only and not because the spray provides a special benefit. One exception— which did receive exemption from the regulations—was a spray used for bronchial asthma attacks. The very fine mist produced by fluorocarbons (which industry was so fond of touting in the case of hair sprays) is extremely important in improving the effectiveness of these sprays. These products were also exempted from the labeling requirement because, the FDA pointed out, "the presence of the warning statement on the label might confuse consumers and dissuade them from purchasing a product that provides a health benefit."

Other products that received exemption from the regulations included contraceptive foams, fixitives used in medical laboratories, a mine-safety warning device, and insect sprays used in aircraft and commercial food-handling areas. These products accounted for about 2 to 3 per cent of total aerosol propellant use in the United States at the time.

The regulations proposed in May 1977 applied only to *propellant* uses of fluorocarbons. (Regulations relating to nonpropellant uses are due this year.) This led to one rather unusual situation. Fluoro-carbons are widely used in the electronics industry as degreasers, solvents, and chilling sprays for trouble-shooting malfunctions. They are used to clean TV tuners, and all military electronic equipment, as well as all Bell telephone switch gears, *must* be serviced with the fluorocarbon sprays. Electronic computers and aircraft navigational equipment are also cleaned with fluorocarbons. According to Dick Pavek, president of Tech Spray, there are no good substitutes for fluorocarbons in these applications. The alternatives are more toxic or flammable, they attack plastic or paint, or they leave a residue that will affect electronic circuitry. At one governmental hearing, Pavek said that some of the alternatives would blow up if sprayed on a hot TV set. Another suggested substitute "will eat up the plastics in a TV set very nicely." According to industry estimates, banning these fluorocarbon products would result in increased annual servicing and maintenance costs for TVs, computers, and aviation and communication equipment amounting to $1 billion or more.

The interesting thing about these products is that, while they are *spray* products and do release fluorocarbons to the atmosphere, the fluorocarbons are not the *propellant*—they are the active ingredient. The propellant, in most cases, is carbon dioxide. Thus, according to a spokesman for the EPA, such sprays will not—at least initially—come under the proposed regulations. Pavek said that, in any event, the electronics industry uses only a small percentage of the fluorocarbons used in the other applications.

The proposed ban on fluorocarbon propellants did not cause those in the aerosol industry to quietly close up shop and steal away —nor should it have. In announcing the regulations, FDA Commissioner Donald Kennedy had pointed out that spray cans using other propellants would not be affected. "We are not acting against aerosol products per se," he emphasized. It was true that the ozone controversy had had some impact on spray-can sales in general, but many in the industry remained convinced that the consumer, given a choice, would not abandon the convenience of the aerosol package.

"The consumer wants aerosols and you can bet the industry will supply them," said David Parker of the Barr Company, an aerosol filler. And if the industry could not supply aerosols with the banned fluorocarbons, they would try to supply aerosols without them. Long before the regulations were announced, in fact, the industry had begun an intensive research program to find suitable alternatives and, at its 1977 midyear meeting, the Chemical Specialties Manufacturers Association announced a big advertising campaign to bring the public back to spray cans. "The days of the dignified soft sell are over," said one CSMA official, apparently oblivious to the fact that spray-can ads have never been notable for their "dignified soft sell." This determination to fight on manifested itself in ads that began appearing on TV late in the summer of 1977, after several marketers had developed aerosols that did not contain fluorocarbon propellants. Alberto VO5, for example, announced that it was going to "challenge the pump."

"But I've got no fluorocarbons," says the pump. "Neither do I," says the new aerosol spray.

Arrid Extra Dry came out with a new product that says "safe for the ozone" right on the can.

The industry wanted the aerosol package to survive, and so there was an intensive search for alternatives to the banned propellants. High on the list of possible substitutes were, surprisingly enough, fluorocarbons. Only a few versions of this family of compounds— principally fluorocarbon-11 (F-11) and fluorocarbon-12 (F-12)—had been banned. However, there are other kinds of fluorocarbons, and several of these contain hydrogen, which makes them susceptible to attack in the lower atmosphere; thus it is likely that only a fraction of these chemicals would ever reach the stratosphere, and they would represent a much smaller threat to ozone than F-11 and F-12. However, these alternate forms of fluorocarbon —precisely *because* they are different from F-11 and F-12—have different chemical and physical properties that make them unsuitable as propellants, or, at least, poor substitutes for the existing ones. Some, for example, are flammable. Moreover, according to David Parker, many are not yet produced in large quantities and are not immediately available even for testing. They might also be two to ten times more expensive than F-11 and F-12. Finally, but perhaps most importantly, the alternative fluorocarbons so far tested have not fared well in animal toxicology tests, according to Ray McCarthy of Du Pont. "Some washed out entirely," he said, while others failed preliminary tests in a way that requires them to be put through further tests for three years. "F-11 and F-12 were so good, you couldn't do anything to any animal with them."

There are, of course, substances other than fluorocarbons that can be, and have been, used as propellants. Although F-11 and F-12 had a virtual stranglehold on the personal-care spray products, carbon dioxide, hydrocarbons, and nitrous oxide (laughing gas) had been used in a variety of products such as shave lathers, dessert toppings, cheese spreads, spray waxes, household cleaners, and many insecticide sprays. (Some spray products used mixtures of fluorocarbons and other propellants.) The possibility of extending these propellants to the spray products that had used F-11 and F-12 was investigated by industry. There was an initial surge of interest in carbon dioxide for cosmetic applications, but, according to Parker, CO_2 ran into problems. Although it is cheaper than fluorocarbons, CO_2 does not produce the proper kind of spray and, because it is not chemically inert, it poses corrosion problems. While CO_2 and nitrous oxide, both compressed gases, worked well in products where a coarse spray is desired (as in automotive products), "Their use in hair sprays has been only marginally successful," Parker said. "Several products have received good initial reception in the marketplace, then have not sustained themselves with repeat sales."

Some marketers have been switching to hydrocarbons instead. These chemicals have been used for many years in some spray products, such as shave foams, but they are now also being used in hair sprays and deodorants. In March 1977, Gillette announced that all fluorocarbon propellants in its products would be phased out by the end of the year, and it introduced several deodorant and hair-spray products containing hydrocarbons. A spokesman for the company said that, while the use of hydrocarbons in such products means there is a "greater flammability consideration," they do not believe there is a "serious flammability risk" if the product is used under normal conditions and as directed. However, the wording on the warning label is slightly different. Users of the old fluorocarbon-containing cans were warned not to spray toward the face or an open flame; users of the new hydrocarbon-containing cans will be warned not to spray near flame "or while smoking."

Virtually up to the last moment before regulations were announced, industry representatives were painting gloomy pictures about the likelihood of finding suitable alternatives. Yet, as events since that announcement have demonstrated, industry can and will come up with alternatives if forced to; in fact, there is good evidence that they knew exactly what they were going to do long before the regulations were announced. What's more, they will peddle the new products with all the missionary zeal they once devoted to defending the unassailable superiority of the old ones. One can only admire their resilience. Despite their complaints that regulations would destroy the industry—surely, by now, almost a Pavlovian response on the part of any industry facing new regulations—it would appear that the aerosol industry is coping nicely, if reluctantly, with the imminent disappearance of fluorocarbon propellants.

If the decision that spray cans by and large represented a nonessential use of fluorocarbons was a value judgment, it was nevertheless one that commanded considerable agreement outside the aerosol industry. But there was even wider agreement that refrigeration *was* an essential use of the chemicals, while air conditioning seemed to be an intermediate category (its importance to any given individual seemingly correlated with the summer conditions in his part of the world).

The industry resisted the idea of sacrificing the propellants to buy time for refrigeration and air conditioning. This was undoubtedly an attempt to save all fluorocarbons by tying their fate to an essential use. But cutting loose the aerosols first was clearly an attractive strategy for the regulatory agencies to adopt when aerosols accounted for such a large proportion of the releases of the fluorocarbons to the atmosphere[12] and you really *could* make a case that stopping these releases would buy time for other, essential uses. John Tukey was so proud of the Academy committee's Solomon-like wisdom in advocating selective regulation, but, in fact, this was not only a sensible, but, indeed, an obvious and unavoidable

[12] The Academy report says that nearly three quarters of the 1975 releases of fluorocarbons came from spray cans. Cooling and refrigeration uses accounted for 14 per cent, with vehicle air conditioners making up 6 per cent and home refrigerators a mere .4 per cent of total fluorocarbon use. The use of fluorocarbons for blowing foam accounted for 12 per cent of total releases.

option once the value judgment had been made that placed spray cans and refrigeration on opposite sides of the essential/nonessential equation. The committee report placed car air conditioners into an intermediate category. "Their usefulness is greater than that of spray cans and less than that of household refrigerators." These air conditioners released about as much fluorocarbons as all the other refrigerating technologies put together, but the report estimated that 90 per cent of these emissions could be prevented by changes in design and service practices or by recovery of the coolant when the car is junked. Regulation should therefore focus "first on service and junking practices . . . and then on redesign." The report suggested that the use of fluorocarbons in home refrigerators might never have to be prohibited.

In 1975, 92 per cent of all refrigerants in use were fluorocarbons, primarily fluorocarbon-12, which was used in domestic refrigerators and food freezers, automobile air conditioners, drinking water coolers, dehumidifiers, and refrigerated food-storage and display cabinets in stores. However, F-11, which was used extensively as a propellant, was not a major refrigerant. Instead, another fluorocarbon, F-22, not a major propellant, was widely used in residential and commercial air-conditioning systems. F-22 was not included in the initial round of regulations. Because it contained hydrogen, it belonged to that group of fluorocarbons considered to be less damaging to ozone than F-11 and F-12, and some people, including Sherry Rowland, suggested that refrigerators be switched to F-22, at least as an interim measure.

However, in congressional testimony in late 1965, Herbert Gilkey of the Air Conditioning and Refrigeration Institute, an industry organization, said that "fluorocarbon refrigerants cannot be used interchangeably." Because of the different properties of the various fluorocarbons, such switches could, he said, result in the inefficient or unsafe operation of the refrigerating system.[13] Gilkey made the case for classifying refrigeration and air conditioning as essential by saying that the fresh-food industry and the chemical, pharmaceutical, aerospace, and optical industries depended on fluorocarbon refrig-

[13] Nor was Gilkey optimistic about the prospects of switching to nonfluorocarbon refrigerants. Not only could the possible alternatives not be used in existing equipment, most of them were either toxic or flammable or both and were not to be used in commercial, residential, public, or institutional applications.

erants; that the communications system, plus all computers and other electronic equipment could not operate without them; and that hospitals also depended on them. Fluorocarbon refrigerants, Gilkey concluded, "are important to all of us, for without them our society could not function."

His contention is hard to dispute, even allowing for industry self-interest and for the fact that some refrigerant and air-conditioning applications could be considered luxuries. (And some practices within the industry could be considered wasteful; for example, it had been the practice not only to empty commercial refrigeration and air-conditioning systems of their coolants periodically, but also to flush the systems out with fluorocarbon cleansers before refilling them.)

While some cutbacks might be made in some applications, the fact remains that a total, sudden ban on fluorocarbon refrigerants would seriously disrupt the life-style of the nation—probably more so than the voters would tolerate. Since one can make a case that this use of fluorocarbons is essential, the prudent course would be to more rigidly enforce containment of the chemicals. Questioned by Senator Dale Bumpers, Gilkey acknowledged that the technology existed for sealing units that in the past had leaked fluorocarbons, but that it "may not be being used as well as it should be used."

The problem of recovering fluorocarbon coolants from junked refrigerating equipment or air-conditioning units would remain. Several people have suggested that laws should be passed requiring such recovery to be made.[14] Industry spokesmen have protested that this would be difficult and expensive, but it remains to be seen how viable this option may be.

One should also, perhaps, be wary of industry claims that there are no alternatives for fluorocarbon refrigerants. This argument has been used too many times, in too many cases, as a bargaining position with which to stall regulation. One suspects that it is not beyond the wit of the chemical industry to come up with substitutes if forced to; after all, Robert Abplanalp *did* announce his new aerosol system the day after the FDA timetable for regulations was released.

[14] F. A. Cotton of Texas A & M University suggested in a letter to *Nature* that an alternative would be to require purchasers of refrigerators to leave a large deposit, returnable with interest, upon presentation of proof that the fluorocarbons had been properly recovered when the unit was junked.

The fate of the fluorocarbon refrigerants will become clearer later this year, when the regulatory agencies will issue their proposed regulations. They are unlikely to deal with the refrigerants as harshly as they did with the propellants; it is probable that the regulations will focus, at least initially, on requirements for better sealing of refrigerating and air-conditioning systems. Yet it is clear that a process has begun that will almost certainly result, in the United States, in strict controls of all releases of fluorocarbons to the atmosphere, if not necessarily a total elimination of their use.

Whether there will be global action on the fluorocarbon problem remains to be seen. The responses around the world have varied widely. The United States and Canada (which announced its intention to regulate fluorocarbon propellants in the fall of 1976) are at one end of the spectrum. In December 1977, they were joined by Sweden, which became the first European country to ban the propellants. The ban, which is to take effect in 1979, was strongly opposed by Sweden's European neighbors, particularly Finland, an aerosol exporter. The European Economic Community recommended a review of the problem in the last half of 1978. The French and the British have come out emphatically against immediate regulation, preferring to await the results of further studies.

During the summer and fall of 1977, new evidence came to light that provided further justification for the action taken against fluorocarbons. As we mentioned earlier, continuing studies of chemical reactions in the stratosphere, and new computer calculations based on those studies, caused the predictions of ozone depletion to be altered substantially—down in the case of the SST and up in the case of fluorocarbons.[15]

This development presented the fluorocarbon industry with some-

[15] This has interesting results. Say that chlorine sources such as fluorocarbons continue to enter the stratosphere. As we have seen, the continued addition of NO_x won't itself have a big ozone depletion effect, but this NO_x will lead to the creation of another chemical that will in turn, lead to an increase in the ozone depletion caused by the *chlorine*. Thus, the NO_x has an indirect effect on ozone. This makes the combination of SSTs and spray cans a bad one—in contrast to the situation we described with respect to chlorine nitrate, where the combination of NO_x and chlorine tended to reduce the ozone depletion.

thing of a dilemma. Recall that throughout the Spray-can War, one of their major arguments was that the atmosphere was a tremendously complicated place and that scientists did not have a completely accurate chemical picture of the stratosphere. The new findings supported this contention, but (and here was industry's dilemma) those findings had the effect of making the fluorocarbon problem much *worse* than Rowland and Molina originally said it would be. Industry could not return to their refrain that scientists lack a complete knowledge of the stratosphere without also calling attention to the fact that their product was even more strongly indicted than before.

It is perhaps not too surprising, then, that industry chose not to argue too strenuously that here was a case of something having been left out of the original calculations.

CHAPTER TWELVE

The Tip of the Iceberg?

List all the substances not yet discovered
that can destroy the ozone layer.

—York University exam question, April 1, 1977.

The question was an April Fool's joke, of course. But it did serve to
illustrate the point that, by early 1977, the list of potential threats to
the ozone layer had grown almost ludicrously long. As Tom
Donahue once said during congressional hearings on the fluorocarbon problem: ". . . we are most conscious, because of the rate at
which our inventory of dangers to the atmosphere has been growing, that we have probably not exhausted the catalogue of horrors."

What are these other "horrors" and how serious a threat do they
pose? They fall into two general categories. The first category includes a group of chemicals similar to fluorocarbons in that they
contain chlorine, which may be liberated to chew up ozone; the second category, which we will discuss in some detail later, includes
nuclear weapons and chemical fertilizers, both of which return us
again to the NO_x problem.

In addition to the chlorine-containing compounds we have already discussed (fluorocarbons, carbon tetrachloride, methyl
chloride), there are a number of others that are also released into
the atmosphere, mostly because of human activities. In fact, the
amount of these chemicals released greatly exceeds that of fluorocarbons,[1] but they have not caused as much concern because they
are removed to a larger extent by chemical reactions in the lower at-

[1] In 1973, some 3 million tons of these nonfluorocarbon compounds were
produced—four times as much as the .75 million tons of F-11 and F-12.

mosphere. However, a certain proportion of these chemicals *does* make it into the stratosphere, and this must be considered in our analysis of the overall threat to the ozone shield.

The amount that does get into the stratosphere differs for each compound. A particularly interesting example of this concerns two chemicals used as industrial cleaning and degreasing agents— trichloroethylene and methyl chloroform. The former is the solvent most widely used for cleaning anything made of metal; much of it gets into the atmosphere by evaporation. However, it reacts chemically quite quickly in the lower atmosphere, which means that it poses little danger to ozone. Unfortunately, the fact that it reacts in the troposphere means that it contributes to urban smog. For this reason, the Environmental Protection Agency has restricted its use. The substitute was methyl chloroform. The EPA said that this was O.K.; since methyl chloroform reacts slowly in the troposphere, it does not contribute to smog. But the very fact that it reacts slowly means that a significant fraction of it will eventually make its way up to the stratosphere. The use of this solvent is growing at a rate of about 20 per cent a year, and by 1979, close to 1 million tons will be produced. McConnell and Schiff calculate that constant production at the estimated 1980 levels would lead to an ozone reduction of about 2 per cent at steady state. We will discuss later why this seemingly small figure of 2 per cent may be cause for concern.

This example is also an excellent illustration of the complexity of environmental pollution problems and the traps that they set for regulatory agencies. All too often, a regulation aimed at solving a particular pollution problem doesn't really solve the problem—it just "exports" it somewhere else and may, in the process, make the problem worse. In this case, the laudable motive of trying to keep the air in cities breathable resulted in transporting the pollution problem to the stratosphere and making it a global, instead of a local, dilemma.

Before returning to NO_x, one other ozone-depleting substance should be mentioned: bromine. In Chapter Eight, we discussed Mike McElroy's suggestion (and the trouble he got into as a result of it) that bromine is so effective at destroying ozone that someone might try to use it as a weapon. It's true that bromine is even more effective than chlorine at chewing up ozone, but it does not appear that very much bromine is released into the atmosphere. At present, most of it seems to come from nature, primarily from the oceans. Methyl bromide is used as an agricultural fumigant, and it is estimated that the annual release of this substance to the atmosphere is about 20 to 40 per cent of the natural production. Its use is growing at about 7 per cent a year.

Bromine is also a component of chemicals (similar to fluorocarbons) that are used as fire retardants in clothing; this use has been growing rapidly. Bromine is also found in gasoline additives.

The chemistry of bromine in the atmosphere is still not well established. It does not appear that, at its present level of production, it will pose a serious problem, but it is one of the many chemicals whose use we must watch.

The ozone controversy abounds in absurdities, but there is perhaps none to equal the part of it that concerns nuclear weapons. In his book *The Genesis Strategy: Climate and Global Survival,* climatologist Stephen Schneider recalls being stunned when he was asked to sit on a National Academy of Sciences panel that would investigate the atmospheric and climatic consequences of a full-scale nuclear war. Schneider was incredulous that "some people actually need to be clubbed with the knowledge that a nuclear war can cause severe global environmental damage before they become convinced that there can be no winners in such a conflict."

As we have mentioned before, nuclear weapons can be a source of stratospheric NO_x. The fireball from a nuclear explosion would produce vast amounts of NO_x from the heated air. We've seen that any combustion engine—in a car or a plane—will convert some of the nitrogen and oxygen in the air to NO_x, and a bomb is a very effective engine in this respect.

John Hampson was really the first to become seriously concerned about the impact of nuclear weapons on the ozone layer. In the late 1950s, he was working at the Canadian Armaments Research and Development Establishment (CARDE) when he was asked to participate with the Americans in studies on how to detect intercontinental ballistic missiles leaving or re-entering the earth's atmosphere. To identify the missiles, it was necessary to study the background of the natural atmosphere.

Hampson's studies led him to recognize the role of catalytic chain reactions in destroying ozone. As we saw in Chapter Two, he was the first to work out the hydrogen (HO_x) chain and was later the first to suggest that water vapor from SSTs might destroy ozone. However, since he worked for a military research establishment, it was natural that Hampson would turn his attention to nuclear weapons. He recognized that the amount of energy required to trigger the destruction of the ozone layer was within human capability to generate and, in particular, was available in the nuclear weapons already stockpiled. He became seriously concerned that a nuclear war would decimate the ozone layer, and he tried desperately to get the Canadian and U.S. military authorities to pay some attention to the problem, but without success. By that time, they had even lost their interest in the kind of atmospheric research Hampson had been doing; it was about this time that the military more or less abandoned the atmospheric sciences.

But Hampson did not give up. The ozone problem began to have a profound effect on him, and he started to brood about the fragility of the ozone layer. He believed, perhaps naïvely, that a recognition of the ozone problem by the major nuclear powers could result in genuine disarmament, and he has tried, almost obsessively, for several years to get people to consider the issue. In a 1974 paper in the British scientific journal *Nature*, he pointed out that military strategists consider the consequences of a pre-emptive nuclear strike only in terms of the direct casualties that might be suffered by either or both of the combatant nations. He argued that they should also consider the possibility that, by unleashing their nuclear arsenals, they might wipe out the ozone layer. This, he said, would be the ultimate "doomsday machine," for it would harm not only the intended victims of the nuclear attack, but the perpetrator as well. It would, in fact, ensure that no one on earth had any chance of escaping the devastation resulting from a nuclear war.

Although Hampson received very little direct credit, things began to happen shortly after his *Nature* paper appeared. In the fall of 1974, Fred Iklé, director of the U. S. Arms Control and Disarmament Agency, gave several speeches in which he emphasized the

hazards to all life on earth that might result from the ozone depletion caused by nuclear war. His remarks received considerable press
coverage and provoked a response from the Pentagon. What was
most disturbing about the Pentagon's reaction, as quoted in the
New York *Times,* is not only that military officials conceded that
perhaps 50 to 75 per cent of the ozone layer might be destroyed, but
also that the Pentagon really didn't seem too worried about this. Incredibly, the officials added that, in any event, the larger Russian
warheads would be responsible for most of the ozone depletion.

The *Times* editorially described the Pentagon's reaction as "altogether preposterous. . . . To plan for a type of war that can expose
the entire world to something far worse than nuclear fallout—that
is folly. It is comparable to a child skipping through a minefield on
the theory that he won't necessarily step on a mine, and if he does,
it won't necessarily prove fatal."

Iklé was hopeful that the ozone connection might be a useful bargaining tool in disarmament talks, and he asked the National Academy of Sciences to do a study.

This was not an in-depth study, as the fluorocarbon investigation
was. The Academy held a five-day workshop in January 1975 and
released a report that summer. This report did not consider casualties from the direct hits of belligerent nations, but the aftermath
effects of the war, particularly on noncombatant nations. Nor did
the study confine itself solely to the ozone question but, as we shall
see, the ozone effects were a prominent feature of the report. In
fact, the Academy's president, Philip Handler, said that the "principal new point" developed in the study was that the ozone effect, not
dispersion of radiation, would be the major impact on countries not
directly involved in the conflict.

The study considered what would happen if ten thousand mega-
tons of nuclear weapons—about half of the then-existing arsenals—
were exploded. The conclusion was that the amount of NO_x in the
stratosphere would increase by factors of from five to fifty (depending on the altitude of detonation and allowing for uncertainties in
the calculations). This in turn would lead to an ozone depletion in
the atmosphere over the Northern Hemisphere of from 30 to 70 per
cent and from 20 to 40 per cent in the Southern Hemisphere. The
peak effect would occur within a few months of the event, and the

atmosphere would take twenty to thirty years to recover. In addition to predicting increases in skin cancer lasting over forty years, the report said that short-term effects would include "severe sunburn in temperate zones and snow blindness in northern latitudes. . . . For a 70 per cent decrease in ozone, severe sunburn involving blistering of the skin would occur in ten minutes."

The Academy report was sent to Iklé with a covering letter from Handler that later became controversial. In the letter, Handler said that a decade or so after the war, areas distant from the conflict would be subject to "relatively minimal stress." The biosphere and man—though not necessarily his civilization—would survive. Handler worried that this "seemingly optimistic assessment" would give "false assurances" to some nations that might consider the advantages of provoking a war between other powers. "Let no reader conclude from this report that distant other nations would survive a major nuclear exchange unscathed and, thereby, inherit the earth." While the physical and biological consequences would be "less prolonged and less severe than many had feared," the economic, social, and political consequences remained "entirely unpredictable."

The press release issued by the Academy focused strongly on Handler's optimism about the survival of the human race and the environment in noncombatant nations. Inevitably, many news stories did the same. But Handler's statement was immediately denounced by the Federation of American Scientists, an activist group, as "bizarre, unnecessary and possibly counterproductive."

Philip Boffey, a writer for *Science* magazine and the author of *The Brain Bank of America,* a book about the Academy, reported shortly afterward that Handler was upset that the Academy "had failed to draw what he considered the most important lesson from the study's own data—namely, that the aftereffects of a nuclear holocaust would be so devastating that there would be 'no hiding place' for anyone. Unfortunately, he said, his letter and the report itself gave the opposite impression."

Why did Handler choose to emphasize the survival of the human race—something he later conceded was his own personal opinion? Boffey wrote: "Handler explained that over the past several years, he has been visited frequently by student activists, and he has repeatedly sought their views on the dangers posed by large nuclear arsenals. He found the students strangely uninterested, apparently because they are paralyzed by the idea that nuclear war would destroy mankind. Thus, to undercut that notion, Handler said, he emphasized that the species would survive." But at the same time, Handler did not want the report to leave the impression that nuclear war was not the hell that it had been assumed to be. His letter was intended to put the report in perspective, but he lamented to Boffey that he "failed to make clear that the report does indeed contain information so disturbing that it should serve as an implicit warning against the hazards of nuclear war."

Is there any evidence that the explosion of nuclear weapons actually *does* have the predicted effect on ozone? Fortunately, we have not yet had a nuclear war to test the point, but there has been atmospheric testing of nuclear weapons. In 1973, H. M. Foley and M. A. Ruderman were the first to suggest that nuclear explosions would inject NO_x into the stratosphere and that a decrease in ozone should have been observed from the testing done by the United States and the Soviet Union in 1961–62. Several models suggested that the decrease might be about 4 or 5 per cent. However, the attempt to analyze the data to see if this happened failed because the ozone measurements in the 1960s were not good enough; there was, then as now, the problem of its large natural variability to contend with. (It was later determined that average global ozone *increased* following the bomb tests, which Hal Johnston explained as the recovery of the atmosphere from the testing. But other scientists believe that this was just part of the natural long-term cycles in ozone.)

Scientists also attempted to look at satellite measurements of ozone taken at the time of the Chinese and French nuclear tests in 1970. They did see a 1 per cent decrease where the explosions occurred, but not anywhere else. Again, the data were right on the edge of the detectability of the predicted effects, so scientists were unable to confirm or refute the predictions on the basis of single test explosions.

Like every other aspect of the ozone debate, predicting the effect of nuclear weapons remains fraught with uncertainties. Two major questions in this case are: How much NO_x is actually made by the bomb? How much of it gets transported into the stratosphere?

At present, the estimates contained in the Academy report constitute the "best guess." Handler, from his special perspective, may have found the report "seemingly optimistic," but in truth it is hard to find optimism in the prospect of 20 to 70 per cent ozone reductions, not to mention the environmental devastation in countries directly hit.

People can agree on the merits of not having a nuclear war, even without the ozone issue being tossed into the argument. The case of fertilizers is a good deal more problematical, for fertilizer use *is* important if we are to feed the world's growing population.

The fertilizer problem has elements in common with both the SST problem and the fluorocarbon problem. Like the SST, fertilizers threaten ozone by contributing to an increase in NO_x in the stratosphere. Like the fluorocarbons, however, this threat originates with a substance released at ground level that gradually makes its way up to the stratosphere. This substance, nitrous oxide (known as laughing gas), is a chemical sometimes used as an anesthetic.

Almost all of NO_x in the stratosphere comes from nitrous oxide, which is released at the earth's surface primarily by biological processes (more on this later). Nitrous oxide is quite inert, it does not dissolve in the oceans, and it is not rained out, so it becomes evenly mixed throughout the troposphere and eventually makes its way up into the stratosphere. There, most of it is split up by the sun's energy to form nitrogen molecules, which are added to those that make up a large part of our atmosphere, but some of it is converted to the form of NO_x that destroys ozone. Since the biological processes that make nitrous oxide have been going on for millions of years, a steady state has been reached—that is, the amount of nitrous oxide being created equals the amount being destroyed—and the natural level of ozone that exists today has adjusted to the continuous production of NO_x resulting from this nitrous oxide.

What, then, is the connection between nitrous oxide and fertilizers? To understand this, it is necessary to review briefly the natural nitrogen cycle on earth.

Nitrogen is an essential component of protein, the most important building material for living tissue. The earth's atmosphere contains about 80 per cent nitrogen, but it is in the form of a stable gas that cannot be directly utilized by living things. This nitrogen must be converted into useful forms—such as ammonia or nitrates—before it can be incorporated into living tissue, and this process of conversion is called *nitrogen fixation*.

Despite the fact that all creatures require nitrogen for survival, few of them can "fix" nitrogen from the air. In nature, most nitrogen fixation is accomplished by certain bacteria that live in a symbiotic relationship with the roots of leguminous plants such as soybeans, clover, peas, and beans. (Neither the plants nor the bacteria can fix nitrogen on their own.) To a lesser extent, blue-green algae also fix nitrogen.

The fixed nitrogen is taken up from the soil by plants, which, in turn, provide protein for animals and humans. It is returned to the soil (or the oceans) through the excreta of animals and, of course, from the remains of their bodies when they die. Some of this nitrogen is recycled through other living beings, some of it finds its way into the deep oceans, and some of it gets tied up in soils, rocks, and sediments, from which it may not escape for perhaps thousands of years. But some of it is converted back to nonfixed nitrogen, through the action of different bacteria in the soil and in the oceans. This process is called *denitrification*.

Denitrification occurs primarily when bacteria have difficulty finding oxygen for respiration; this usually occurs when they're in water or wet soils. Being more versatile than humans—who would give up and die under the circumstances—the bacteria simply switch to using nitrates as their supply of oxygen. In the process, they convert the nitrates to nitrogen gas and, to a lesser extent (about 6 per cent), to nitrous oxide, both of which are released to the atmosphere.

Thus it can be seen that the amount of nitrous oxide produced depends on the amount of fixed nitrogen in the soil. Fertilizers are a form of fixed nitrogen; when we add them to the soil, we increase the rate at which nitrous oxide is formed, and this will eventually have an impact on the ozone layer. How large an impact this is

likely to be is a subject of hot debate. Here scientists are in even
worse shape than they were in trying to predict the effects of
fluorocarbons; the fertilizer problem is even more mind-boggling in
its complexity. The fluorocarbon problem involved questions and
uncertainties related to air motions and chemistry; the fertilizer
problem involves these uncertainties, plus others associated with
elaborate biological cycles on land and in the oceans. (One of the
more intriguing aspects of the fertilizer controversy was the sight of
meteorologists and atmospheric physicists and chemists desperately
boning up on oceanography, microbiology, and soil and agricultural
sciences.)

Predicting the amount of ozone depletion that could result from
the use of fertilizers depends on several critical factors, all of which
are subject to considerable controversy. The first is the amount of
nitrous oxide that is produced in denitrification, relative to the
amount of ordinary nitrogen gas produced. (Remember that the
bacteria produce mostly plain nitrogen gas, which is the same as 80
per cent of the atmosphere and is harmless to ozone.) In terrestrial
soils, the ratio between these two substances depends critically on a
number of factors, including the moisture content, the availability
of oxygen, and the acidity of the soil. (Acidity increases the ratio of
nitrous oxide produced. Human activities have been adding mainly
acidic pollutants, such as sulphur dioxide, to the soil.)

The second important question is: How long does denitrification
take? In other words, what is the time span from the point at which
nitrogen is fixed to the point where it is converted to nitrogen gas
and nitrous oxide? For example, if we were to add fertilizer to the
soil, some would be denitrified in a matter of weeks. Depending on
the conditions mentioned above, the amount denitrified could range
from 1 to 75 per cent.

Some of the fixed nitrogen in the fertilizer is taken up by annual
crops, which are then eaten by animals or man. Some of the nitrogen returns to the soil quickly through the excreta of animals. Some
of it becomes incorporated into the animals' bodies and is not returned until the animals die; thus it may not be recycled for several
years or decades. The nitrogen taken up by trees will also be tied up
for many decades. Some of the fixed nitrogen runs off the land and

ends up in the deep ocean, where it becomes tied up in the sediments or in inorganic substances. It could take a thousand years for this nitrogen to be denitrified. The oceans and their sediments are, in fact, the greatest reservoirs of fixed nitrogen, containing about a thousand times more than any of the other reservoirs we have mentioned.

The upshot of all this is that the length of time required for denitrification to occur is the largest uncertainty affecting calculations of the impact of fertilizers on the ozone layer. Another major uncertainty concerns the role of the oceans. Some scientists believe the oceans are a major *source* of nitrous oxide. Others—notably Mike McElroy—do not accept this view, and this in large part accounts for the large differences in the predictions of ozone depletion. This factor is extremely important in calculating how much impact fertilizer use will have. Man's intervention occurs mainly on land; if the major source of nitrous oxide is the ocean, then our activities may be contributing only a small percentage of the total amount of nitrous oxide, and our influence may not be that important. However, the question of whether the ocean is a large source of nitrous oxide remains unresolved.

Despite the uncertainties, however, there is still cause for concern about our use of fertilizers. The Council for Agricultural Science and Technology has suggested that a major agricultural revolution might pose more of a threat to ozone than aircraft or industrialization would.

One of the major factors contributing to this concern has been the dramatic increase in fertilizer use in recent years—the result of our efforts to increase agricultural productivity and feed the world's ever-growing population. (Fertilizer use on genetically improved "miracle" grains was an essential strategy in the "Green Revolution" that resulted in increased yields in many Third World countries.) Early in the debate, Tom Donahue cautioned that the ozone problem might put "a natural ceiling" on the already grim prospects for preventing much of the world's population from starving.

The total fixation of nitrogen by all means is estimated to be about three hundred million metric tons a year. In 1950, man-made fertilizer contributed about 1 per cent of this. By 1974, it had grown to 15 per cent, and it is estimated that by the turn of the century we will be fixing nitrogen at a rate comparable to or greater than the natural-fixation processes.[2] (Fertilizer use has been growing at a rate of about 9 per cent a year, but it should be noted that very large amounts of increasingly expensive energy are needed to produce fertilizers, and this may have a braking effect on the growth in its use.)

What are the predictions of the impact this fertilizer use might have on the ozone layer? Most of the modelers were able to give predictions for the impact of the SST and the fluorocarbons that were in reasonable agreement, but this is not the case for fertilizers. The model predictions differ widely.

There are two aspects to the predictions. The first concerns the long-term steady state ozone depletion, which would eventually occur if fertilizer use was leveled off at a given time and remained constant thereafter. The predictions here have ranged from 12 to 50 per cent ozone reduction; the difference is due in large part to differing opinions about the role played by the oceans.

But there is an even greater uncertainty—and greater disagreement—about the time it will take to reach steady state. This depends on the length of time it takes for denitrification to occur, a factor, we have seen, that is highly uncertain. McElroy estimated that a 20 per cent ozone reduction by the year 2025 is not unlikely. At the other extreme, Donahue and Liu said that the deep ocean sediments and soil minerals on land tie up the fixed nitrogen for so long—and that this is such a large reservoir of fixed nitrogen—that this dominates the time scale for denitrification. Human activity is puny by comparison and we may not reach steady state, in terms of our contribution, for hundreds of years. Hal Johnston took a somewhat similar view. He said the steady state ozone reduction could reach 10 per cent, but that could take several hundred years. However, he said there could be a 1 to 2 per cent reduction before the end of the century.

The fertilizer problem, already scientifically messy, has become

[2] Combustion is another human activity that produces fixed nitrogen, but this factor is not important in the ozone question because the NO_x in this case is rained out. The same is true for fixed nitrogen produced naturally by lightning.

even messier as a result of new measurements—the same ones that brought down the predictions of SST impact on ozone. It is too early to tell how large an effect this will have on the predicted impact of fertilizers, although it seems clear that it will, as in the SST case, lessen the potential effect. This does not mean the fertilizer problem is unimportant, but it does mean that it appears to be less urgent and that we have more time to study the issue than we did for fluorocarbons, which is lucky, since the fertilizer problem is without question infinitely more complex.

The seemingly small figure of 1 to 2 per cent ozone depletion by the turn of the century can, however, be misleading. Remember that there are multiple threats to the ozone layer and that *their individual effects are largely additive*. Fertilizers may cause only a 1 to 2 per cent ozone depletion by the year 2000, but we also continue to use fluorocarbon refrigerants, methyl chloroform, and other chlorine-containing compounds as well. The space shuttle is scheduled to go into operation in the 1980s. If all of these things add a per cent here and a per cent there, it's not long before you're talking about a total ozone depletion of 10 per cent or more. We could be wiped out not by a single large technological goof, but by the sum of many small ones. And the main fear comes not from the threats we have already identified, but from those we have not. As McElroy once said in congressional testimony: "What the hell else has slipped by?" If being asked to "list all the substances not yet discovered that can destroy the ozone layer" sounds absurd, it is nevertheless essentially what we must do.

All of this poses a nightmarish dilemma for the legislators and regulatory agencies whose job it is to ensure that the environment is protected from pollution. How do you regulate widely diverse technologies that each cause a small amount of ozone depletion? The obvious answer is to put an upper limit on the total amount of ozone depletion that will be tolerated and apportion it among the various technologies. Even assuming that this could be calculated with precision, how would you decide just how much ozone depletion each technology would be allowed to cause? The technique of doing a cost/benefit analysis comes to mind. But who is to make the judgment on benefits? And how do you decide between the relative benefits of, say, fertilizer to grow more food or refrigerator cars to transport that food to major cities? Of course, the situation becomes infinitely more complex at the international level. Even supposing that international agreements on restricting ozone-depleting technologies could in principle be achieved—by no means a foregone conclusion, if the fluorocarbon issue is any example—are different countries likely to make the same value judgments about the relative merits of the different technologies? Not very likely. For example, increasing fertilizer usage is likely to be considered more important in Third World countries, where it is necessary for the success of the "miracle" crops, than it is in the United States, where usage is already so high that increases soon hit the point of diminishing returns. (In the United States, incidentally, some 10 per cent of the fertilizers used goes on golf courses.) On the other hand, food distribution in this country depends critically on advanced refrigerated transportation and storage; the major cities would not last more than a few days without it. This problem might not be so acute in many other countries.

Another factor to consider is that many developing nations simply are not as concerned about pollution as the developed nations are; they are more interested in encouraging the industrial development that will increase their standard of living.

And what about the space shuttle? Will the rest of the world look favorably on letting it chew up its little bit of ozone if this means giving up the benefits of some other technology?

These questions are many and complex and we have barely begun to think about them, much less answer them. One thing is certain, however: Despite the fact that it took four years to decide what to do about fluorocarbon propellants, this was probably the easiest ozone problem we have been asked to solve. The hard ones are yet to come.

What the ozone controversy has demonstrated is that life on this planet depends for its existence on a very small amount of a very unstable and easily destroyed substance. Ozone is the weakest link in the earth's life-support system. If that seems to you a precarious gd conclusion

state of affairs—it is. If we break the chain, life on this planet may have to wait until evolution can cope with the changed conditions. If we are careless enough to let this happen, it is to be hoped that whatever creature may evolve as a result will be capable of learning from the many and grievous mistakes of the human dinosaur.

Guilty Until Proven Innocent?

I believe firmly that we cannot afford to give chemicals the same constitutional rights that we enjoy under the law. Chemicals are not innocent until proven guilty.

—RUSSELL PETERSON, former head of the Council on Environmental Quality.

The fluorocarbon controversy was more than just a dispute over whether certain kinds of chemicals should be used as spray-can propellants. It was a particularly illuminating example of an increasingly common dilemma that faces society today: How do we cope with threats to the environment whose effects we cannot predict with certainty, but whose consequences may be extremely serious? There would be little problem if we could be certain of the consequences, but, as the preceding chapters have demonstrated, science cannot give us a simple yes-or-no answer—on the fluorocarbon problem or on any number of equally complex social problems resulting from our reliance on technology.

Of course, lack of certainty cannot be translated into lack of action—not anymore. Maintaining the status quo *is* doing something. It is a decision by default, but a decision nonetheless. So if we *must* decide in the face of substantial scientific uncertainty, we can turn to a method we understand, one that has served us well: The trial.

The environmental trial is still in nascent form. The rules of the game have not been entirely worked out and the process is not totally formalized, as it is in a real court of law. The roles of judge, jury, and attorneys—and who should play them—have not been clearly spelled out. Nevertheless, the fluorocarbon debate had many elements of such a trial, and there was none so pervasive as the question of burden of proof.[1] Here again is a question that applies to a much broader range of issues, but the fluorocarbon debate provides an enlightening case study.

When you ask the question, Who has the burden of proof?, you are really asking, Whose job is it to persuade the judge that his side is correct? In the fluorocarbon case, the question can be put thus: Should industry be allowed to continue producing fluorocarbons until it is proved that the chemicals are, in fact, destroying ozone? Or should production and use of the chemicals be banned until industry can prove that they are not destroying ozone? The issue is more than mere philosophical debate for, should the balance scales turn out to be very nearly equal—should neither side be able to gain a clear victory—then *the one who has the burden of proof loses.*

Industry spokesmen were fond of comparing the "trial" of fluorocarbons to that of a human being in a criminal trial. They point out that, in criminal trials, the accused is considered "innocent until proven guilty," and they protested, often in tones of righteous indignation, that the rules should be no different for fluorocarbons. "Innocent until proven guilty," they asserted, was a fundamental principle of our legal system, and they sometimes acted as though the underpinnings of justice were being menaced and even subverted by the suggestion that perhaps it was up to them to provide some reasonable assurances that the chemicals they manufactured were not harming the ozone layer that belongs to us all.

One can question the analogy with the criminal trial. The fluorocarbon dispute can be more appropriately likened to a civil trial and, in civil cases, the premise "innocent until proven guilty" does not unfailingly apply. But even allowing the analogy with the criminal trial, one can question whether the underpinnings of justice were indeed being menaced by the suggestion that the burden of proof rested with industry. It is surprising to discover that the

[1] The following discussion of burden of proof was derived from an interview with Neil Brooks, law professor at Osgoode Law School, York University, and from a speech on the legal implications of genetic engineering given at York by A. M. Capron of the University of Pennsylvania Law School. We are especially grateful to Brooks for drawing our attention to the legal and philosophical premise underlying the concept of "innocent until proven guilty"—something that the public may not generally be aware of.

underlying principle in the criminal trial is *not* "innocent until proven guilty." This premise is the *result* of a recognition that the law is an imperfect, uncertain instrument—that mistakes can be made. So the real question is this: If we make a mistake, which error would have the more serious consequences—which would carry the greater risk for society? Specifically, would it be a greater error to allow a guilty person to go free or to convict an innocent person? As a society, we have made the value judgment that the greater risk and the most serious consequences would result from making the error of convicting an innocent person. (Hence the concept, "Better to let ten guilty men go free than to convict one innocent man." However, with increasing crime rates, society's values on this may well change.) The result of this is that the state bears the burden of proving guilt, and the accused is therefore considered innocent until proven guilty.

Let us apply this same reasoning to the fluorocarbon case. There are two types of errors that could be made here. The first is that we will judge the ozone layer to be in danger when in fact it is not. The other is that we will judge the ozone layer not to be in danger when in fact it is. Which of these errors carries the greater risk for society as a whole? Clearly, the second. And this, of course, is precisely the error we stand to make if we adopt industry's wait-and-see position. Thus the burden of proof rests with industry; they must show that the chemicals are not harming the atmosphere.

Having arrived at that conclusion, it is necessary to ask what standard of proof is required. In a criminal trial, because the consequences are so serious and because criminal sanctions will apply, the state must prove its case *beyond a reasonable doubt*. In civil cases, the consequences are usually deemed to be less serious, and the standard is lower—it is *preponderance of evidence* or the *balance of probability*.[2] Since the procedures of an environmental trial have

[2] Hal Johnston, in a letter to *Science*, once outlined the SST controversy in legal terms: In 1971, there was sufficient scientific evidence to establish *probable cause* that NO_x from SSTs flying at twenty kilometers would reduce ozone (probable cause being enough for "a grand jury to recommend that a case be tried in a court of law rather than be dismissed"); by 1976, there was sufficient scientific evidence to *prove beyond a reasonable doubt* that SSTs would reduce ozone, and, finally, that there is a *preponderance of evidence* that reductions would cause increases in skin cancer unless people moved or changed their life-styles (emphasis Johnston's).

not yet been precisely formalized, it is not clear which standard would apply. Of course, in an environmental trial, no criminal sanctions are applied, and so the less stringent standards of the civil trial would seem to be appropriate.

Regulatory officials concerned with the fluorocarbon issue very clearly adopted the position that the burden of proof rested with industry. For example, Wilson Talley, assistant administrator of the Environmental Protection Agency said, at the Utah meeting just two days after the release of the National Academy's fluorocarbon report, that it is not acceptable to postpone decisions indefinitely waiting for better data to come in. And demanding scientific certainty before acting would mean waiting for actual "body counts" to prove that a danger exists. "I am convinced that the public interest demands precautionary environmental regulations, based on the best data available, early enough to assure that no such 'body counts' are ever needed."

There were two other aspects of the fluorocarbon issue that made the regulatory problems even more difficult. They are not unique to this controversy, but, again, the fluorocarbon case provides a good illustration. The first problem has to do with regulating activities whose harmful effects are delayed for a considerable period of time; the second, a related issue, is the problem of obtaining informed consent from those who must assume the risks.

As we have seen, the full ozone depletion effects of fluorocarbons already in the atmosphere have not yet occurred. The impact to date is below measureable limits. Moreover, once the impact *does* become measureable—which would be the proof that the theory is correct—the situation must necessarily get worse for some considerable time thereafter. Thus, said Russell Peterson, former head of the Council on Environmental Quality, "the decisionmaker must—as soon as he has reasonable assurance that the predicted effects will occur—consider the potential future effects as if they were taking place in the present. . . . The decisionmaker may not be able to wait for a measurement of the effects."

The issue of informed consent is problematical in this context. The concept arose first in connection with scientific and medical experiments requiring the use of human subjects. Such experiments are now rigorously controlled, and researchers wishing to use human subjects must demonstrate that this is vital to the experiment and ensure that the subjects have been fully informed of the risks they face and the possible benefits that might accrue to themselves or society as a result of their participation in the experiments. This requirement is grounded in the philosophy that those who are at risk must *understand the nature of that risk and voluntarily assume it.* It is this principle that the aerosol industry so blithely ignored in arguing that an ozone depletion of one half of 1 per cent entailed no greater risk than moving thirty-five miles toward the equator. Moving thirty-five miles toward the equator is a voluntarily assumed risk; living with a global reduction in ozone of one half of 1 per cent is not.

Science can help us to understand the nature of the risks we face, but it cannot tell us whether or not we should assume them. This is a value judgment, and it depends critically on the extent to which we are gamblers, both individually and collectively. It is certainly true, as many in industry frequently pointed out, that we live every day with risks that we have not voluntarily assumed. It is also true that we cannot have a zero-risk world. But it is clear that, in this country at least, there is a rising public militance about having a greater personal say in the nature and degree of risks that we, as a society, are forced to accept—particularly when only a small group of those who bear the risks stand to profit from it. A reader of *Business Week*, L. A. Freeman, once put it this way: ". . . no one expects to live a life completely free of risk. But we all have a right to expect protection against our being involuntarily used to generate private gains for others at the unknowing risk of our lives."

And what of the informed consent of those not yet born? The full environmental effects of fluorocarbons may not be felt for decades and may continue for a century or more. Thus many of those who will be affected cannot even choose to take the risk, as people can today by choosing to use fluorocarbon-containing spray cans. (The problem of storing long-lived radioactive wastes confronts us with a similar ethical dilemma.) But we can no longer ignore the rights of future generations if there is a possibility that our actions may make large parts of the world uninhabitable or very nearly so. More than ever before, an ethic of stewardship of the earth's resources—and, especially, of its ability to sustain life—must govern the decisions we make today about potential environmental hazards of the future.

Some in industry chose not to express the burden of proof question in "innocent until proven guilty" terms. They put the issue this way: Rowland and Molina have presented a scientific hypothesis. The burden of proof is on them to show the validity of their hypothesis, using the normal scientific method. This was indeed a legitimate point. Who could argue that Rowland and Molina—and, by extension, the entire scientific community—did not have a responsibility to verify the theory? That is what science is all about. Of course, few people advocated otherwise. Certainly, Sherry Rowland accepted this, although he did point out that industry's contention was "just hypothesis too. They have the hypothesis that it is safe to release fluorocarbons, but no data to back up their position. We have a hypothesis that it is unsafe, but we do have some scientific data and are coming up with more." Nevertheless, it would be patently unfair to allow mere accusation to shut down an entire industry. There are always environmental extremists who are simply anti-industry, but responsible participants in the fluorocarbon debate who felt that the burden of proof ultimately rested with industry were *not* calling for lynch mob tactics—conviction and execution without trial. What they were saying was that, if the trial did not provide fluorocarbons with a clear-cut vindication, then the chemicals had to be banned.

Of course, scientists who raise such alarms are not without dilemmas themselves. They know that, no matter how much research they do, they probably do not have a complete picture of the problem and they are acutely aware that something can always come along to change that picture completely. This is particularly true when a system as complex as the earth's atmosphere is involved. In 1930, four chemical reactions considered to be important in the upper atmosphere were known; today, there are more than a hundred. With research, scientists may narrow the uncertainties associated with each one of these reactions, but since the total number of processes required for an understanding of the system has increased, the total uncertainty has likewise increased.

What then does a scientist say when he is asked by the public or political leaders for an assessment of the problem? He must answer on the basis of the best available information, but must be careful also to give the range of uncertainty. Many scientists have found, however, that this may still get them into trouble, for the public and political leaders often seem to disregard the caveats about uncertainties and focus instead on the central figure, which is taken as gospel. When new data subsequently shift this central figure (even though the shift may be well within the range of uncertainty given in the first instance) the credibility of scientists is damaged.

All of this poses a problem: If the best current knowledge indicates that a certain technology will cause an environmental threat, should scientists "blow the whistle" on it, even if their knowledge is not complete? If uncertainties remain, should scientists suggest that a technological development like the advanced SST be aborted, or that a commercial enterprise such as spray cans be abolished?

While this is never an easy decision, it should always be remembered that the uncertainties always cut both ways. One must not jump to the conclusion that resolving the uncertainties will necessarily cause the problem to go away and that scientists are therefore alarmists if they call attention to the problem before those uncertainties are resolved. It is *equally probable* that, when the uncertainties are narrowed by further research, the problem will turn out to be worse than originally predicted.

It's somewhat like the probabilities involved in tossing a coin; there is always a 50 per cent chance that the toss will come up heads and a 50 per cent chance it will come up tails. If you take a series of measurements in the laboratory, and give the average as your "best guess," the chance that the next measurement will be higher is equal to the chance that it will be lower.

In one respect, the ozone controversy provided an excellent illustration of this concept. Remember that, in mid-1977, new measurements and calculations resulted in a significant reduction in the predicted impact of SSTs, while *at the same time* resulting in a significant increase in the predicted impact of fluorocarbons. The fluorocarbon case alone is also a classic example. New measurements carried the estimates back and forth several times and this may continue to occur in the future. However, the SST case was a peculiar anomaly in that every new measurement reduced the predicted impact on ozone. It was like tossing fifty heads in a row.

Such an event, unexpected as it may be, was bound to provoke a chorus of I-told-you-so's from SST proponents. It can be argued, of course, that pronouncing the SST safe *before* a study is done on the problem is quite a different matter from doing so *afterward,* but there is, nevertheless, a certain touch of defensiveness in having to explain why the early predictions were, apparently, so wrong. Since that explanation, as described earlier, may be difficult for the non-scientist to accept, it would not be entirely surprising if scientists became increasingly tempted to duck the responsibility of giving political advice on issues like the ozone controversy. But their advice—however two-armed it may have to be—is needed and will probably be demanded by society. It would appear that scientists really have no choice but to offer that advice on the basis of the best information available at the time, with all the pitfalls that entails.

The fluorocarbon issue was frequently mentioned as a possible case to be tried by a formal Science Court. The concept has been bandied around from time to time but in its most recent incarnation, it has been most vigorously pushed by Arthur Kantrowitz, chairman of Avco Everett Research Laboratory Inc. In 1976, Kantrowitz headed a task force of a presidential advisory group that studied the Science Court idea.

According to the task force report, the Science Court would work like this: Once the issue was chosen, "case managers"—people scientifically qualified to argue opposite sides of the case—would be chosen. Essentially, they would be prosecuting and defense attorneys. Judges would be chosen from a list of "unusually capable scientists having no obvious connection to the disputed issue." The judges must be accepted as qualified and impartial by both case managers. In addition, there would be a referee to ensure that proper procedures would be followed in the trial. Once this was done, the two case managers would each prepare statements of scientific fact about the issue; speculation and "iffy" statements would be forbidden. These statements would then be exchanged, and each side could challenge the other's statement. Challenged statements would be handled by an adversary procedure; the case managers would cross-examine each other before the judges. The judges would then issue an opinion on the validity of the disputed statements, outlining the margins of error if necessary, and this, together with the undisputed statements of fact, would comprise the court's report. The court would deal *only* with questions of scientific fact; it would not consider social, political, or economic factors, and it would not make value judgments or recommendations as to what society *should* do about the problem. In other words, it would do very much what the National Academy panel did in assessing the scientific data and uncertainties (although the panel went a step further and did a lot of independent work on the problem), but it would not do what the Academy committee did in recommending a delay in regulations.

We do not propose to discuss the pros and cons of the Science Court here, but simply to pose some questions about the concept that arise from the experience of dealing with the fluorocarbon issue.

One fundamental question is whether the adversary process, in its legal sense, is appropriate to the resolution of scientific issues. As the National Academy's president, Philip Handler, remarked in a New York *Times* magazine article, a Science Court must be "absolutely free of the chance of some Perry Mason-like figure getting a chemical 'off the hook.'" A related question is whether the procedures of the courtroom would be acceptable to scientists. In some trials, particularly civil ones, a lawyer will, if he can, prevent information damaging to his client from coming to light; he will certainly not volunteer such information himself. Contrast this with Sherry Rowland's revelation of the chlorine nitrate factor, despite the fact that, at least initially, it was damaging to the case he was trying to make for immediate regulation of fluorocarbons. It seems clear that scientists would never sanction an adversary procedure that condones suppression of scientific data as a means of winning over a jury; this is contrary to their training and to the ethics of science.

A second question is whether the judge and jury in a Science Court should be composed of scientists who have "no obvious connection to the disputed issue." While this is clearly imperative in, say, a murder trial, it is not obvious that this would be the best course to follow in an environmental trial. It would, of course, be necessary to avoid choosing scientists with an obvious vested interest (Sherry Rowland or Ray McCarthy, for example), but the choice of totally uninvolved scientists might also be a serious deficiency. Scientists without any direct involvement in or knowledge of the relevant fields of science are reduced to sifting through the arguments of adversaries, and this simply increases the likelihood that judgments will be based in large part on the persuasive rhetoric and debating skills of the adversaries, rather than on the scientific facts. Whether scientists like the image or not, they do count "Perry Mason-like figures" among their numbers.

In choosing its fluorocarbon panel, the National Academy opted for a middle route. Half of the members were scientists who were respected in their own fields but whose work had not been directly related to stratospheric chemistry or the ozone problem. The other half were scientists whose previous research was directly related to the ozone problem, but who had not publicly committed themselves to a political position regarding the regulation of fluorocarbons. This is a model that a future Science Court might well emulate.

The final and perhaps most fundamental question regarding the Science Court is whether it is even needed. Do we really need another, inevitably bureaucratic, institution to help us solve these problems? Will the congressional hearings and the National Academy studies—not to mention those done by other scientific organizations and the regulatory agencies—not suffice? If not, can the existing institutions be modified or changed to meet this need more effectively?

We do not have the answers to these questions. What is clear, however, is that society must begin to question whether existing social and political institutions are adequate to cope with a new breed of environmental dilemma. The fluorocarbon problem is just one example; other examples—genetic engineering, nuclear power— already abound, and there seems little doubt that the future holds many more. How *will* we decide what to do when confronted with uncertain hazards? How much are we willing to gamble?

INDEX